SpringerBriefs in Applied Sciences and Technology

For further volumes:
http://www.springer.com/series/8884

U.S. Dixit · D.K. Sarma · J. Paulo Davim

Environmentally Friendly Machining

 Springer

U.S. Dixit
Department of Mechanical Engineering
Indian Institute of Technology Guwahati
Guwahati, Assam, India
uday@iitg.ernet.in

D.K. Sarma
Department of Mechanical Engineering
Indian Institute of Technology Guwahati
Guwahati, Assam, India
dks@iitg.ernet.in

J. Paulo Davim
Department of Mechanical Engineering
University of Aveiro
Aveiro, Portugal
pdavim@ua.pt

ISSN 2191-530X   e-ISSN 2191-5318
ISBN 978-1-4614-2307-2   e-ISBN 978-1-4614-2308-9
DOI 10.1007/978-1-4614-2308-9
Springer New York Dordrecht Heidelberg London

Library of Congress Control Number: 2011944831

Printed on acid-free paper

Springer is part of Springer Science+Business Media (www.springer.com)

Preface

The environment is closely linked with human civilization. In order to ensure preservation of the human race, a healthy environment must be maintained on our planet. There was a time, when the mankind was at the mercy of the whims, fancies, and benevolence of nature, but slowly we started conquering nature, at least partially. Nowadays, we are no less dependent on technology than nature. Technology has made lifestyles more comfortable, but it is putting a heavy tax on the environment. It is being realized that if proper attention is not paid to the environment, we will have to face a lot of health and survival problems. Hence, increasing effort is being put for developing green or environmentally friendly technologies to reduce the negative impact of technology on the environment.

Environmentally friendly machining is a part of green manufacturing. Machining is one of the widely used manufacturing processes, and it is evident that it has to be made environmentally friendly. This book reviews the major efforts made by the researchers to develop environmentally friendly machining. We have deliberately discussed compressed air-cooled machining in somewhat greater detail, because compressed air-cooling can be very easily implemented on the shop floor. However, the other green technologies discussed in the book also have immense potential and they should be explored and used in practice. The last chapter of the SpringerBrief provides directions for future research and development.

This book can be used as a part of a course on machining at senior under graduate and post graduate level. It can also form a textbook for an elective course on environmentally friendly machining. Practicing machine shop engineers and managers will also find it useful.

We thank Mr. Tomi Ado, M.Tech. student at IIT Guwahati, for drawing some figures for the book. We thank Dr. Alison Waldron, Senior Editor Springer, and Editorial Assistant Ms. Merry Stuber for motivating us to work harder and providing all the assistance to enable us to complete this project in a timely manner. We request all our valued readers to provide feedback about the book via e-mail.

Guwahati, Assam, India   U.S. Dixit
Guwahati, Assam, India   D.K. Sarma
Aveiro, Portugal   J. Paulo Davim

Contents

Chapter 1
Green Manufacturing

Abstract Green engineering is used to indicate environmental concerns in engineering. Green manufacturing is a subset of green engineering. Environmentally friendly machining is a part of green manufacturing. It is included in the concept of sustainable manufacturing, which considers economical and social concerns in addition to environmental concerns. This monograph focuses on environmentally friendly machining. An environmentally friendly machining attempts to minimize the consumption of cutting fluid, cutting tools, and energy.

Keywords Green engineering • Green manufacturing • Green machining • Environmentally friendly machining • Sustainable manufacturing • Environmentally conscious manufacturing • ISO 14000 • OHSAS 18001

1.1 Introduction

This monograph deals with environmentally friendly machining. Environmental concerns are gaining importance in every field of engineering. After the ISO 9000 Quality Management System Standards, the ISO 14000 Environmental Management System Standards and the OHSAS 18001 Occupational Health and Safety Assessment Series were published. The ISO 14000 is a set of standards concerning a way an organization's activities affect the environment throughout the life of its products. These activities range from production to ultimate disposal of the product. It includes effects on the environment such as pollution, waste generation, noise, depletion of natural resources, and energy use. The ISO 14000 standards are designed to cover environmental management systems, environmental auditing, environmental performance evaluation, environmental labeling, and life-cycle assessment. The ISO 14000 specifies that an environmental policy, fully supported by senior management, must exist. It outlines the policies of the company, not only toward the staff but also toward the public. OHSAS 18001 is a document for health

U.S. Dixit et al., *Environmentally Friendly Machining*, SpringerBriefs
in Applied Sciences and Technology, DOI 10.1007/978-1-4614-2308-9_1,
© Springer Science+Business Media, LLC 2012

and safety management systems. It is intended to help an organization to control occupational health and safety risks. It was developed in response to widespread demand for a recognized standard against which health and safety measures can be certified and assessed.

At many places, the word "green" is considered as a synonym of "environmentally friendly." This may be due to the fact that most of the plants are green and convert CO_2 to O_2 needed for the survival of human being. Green color symbols are often used to represent plants and vegetables. This book could also have been named as Green Machining, but this term is in use in a different sense. The machining of ceramic in the unbaked (pre-sintered) sate is called green machining. However, green engineering and green manufacturing are words in vogue to indicate the environmental concerns in engineering and manufacturing, respectively.

Manufacturing includes all steps necessary to convert raw materials, components, or parts into finished goods that meet a customer's expectations or specifications. Machining is one type of manufacturing. This chapter gives a brief introduction of environmentally friendly manufacturing in general, and environmentally friendly machining in particular. All other chapters will focus only on environmentally friendly machining.

1.2   Green Engineering

According to the US Environmental Protection Agency, green engineering is the design, commercialization, and use of processes and products which are feasible and economical while minimizing (a) generation of pollution at the source and (b) risk to human health and the environment. In the conference of "Green Engineering: Defining the Principles" held in Sandestin, Florida, in May 2003, more than 65 engineers and scientists developed the Principles of Green Engineering, which are as follows (Abraham and Nguyen 2003):

1. Engineer processes and products holistically, use systems analysis, and integrate environmental impact assessment tools.
2. Conserve and improve natural ecosystems while protecting human health and well-being.
3. Use life-cycle thinking in all engineering activities.
4. Ensure that all material and energy inputs and outputs are as inherently safe and benign as possible.
5. Minimize depletion of natural resources.
6. Strive to prevent waste.
7. Develop and apply engineering solutions, while being cognizant of local geography, aspirations, and cultures.
8. Create engineering solutions beyond current or dominant technologies; improve, innovate, and invent (technologies) to achieve sustainability.

9. Actively engage communities and stakeholders in the development of engineering solutions.

Green engineering embraces the concept to protect human health and the environment with cost effectiveness when applied early to the design and development of a product. Anastas and Zimmerman (2003) have presented the following "12 principles of Green Engineering" [reprinted with permission from Anastas and Zimmerman (2003). Copyright (2003) American Chemical Society]:

Principle 1: Designers need to strive to ensure that all material and energy inputs and outputs are as inherently non-hazardous as possible.

Principle 2: It is better to prevent waste than to treat or clean up waste after it is formed.

Principle 3: Separation and purification operations should be designed to minimize energy consumption and materials use.

Principle 4: Products, processes, and systems should be designed to maximize mass, energy, space, and time efficiency.

Principle 5: Products, processes, and systems should be "output pulled" rather than "input pushed" through the use of energy and materials.

Principle 6: Embedded entropy and complexity must be viewed as an investment when making design choices on recycle, reuse, or beneficial disposition.

Principle 7: Targeted durability, not immortality, should be a design goal.

Principle 8: Design for unnecessary capacity or capability (e.g., "one size fits all") solutions should be considered a design flaw.

Principle 9: Material diversity in multicomponent products should be minimized to promote disassembly and value retention.

Principle 10: Design of products, processes, and systems must include integration and interconnectivity with available energy and materials flows.

Principle 11: Products, processes, and systems should be designed for performance in a commercial "after life."

Principle 12: Material and energy inputs should be renewable rather than "depleting."

Green manufacturing is a subset of green engineering. Most of the principles of green engineering are valid for green manufacturing too. However, sustainability is a more general term than green engineering. It considers economic, social, and environmental issues together (Toakley and Aroni 1998).

1.3 What is Sustainable Manufacturing?

Recently, the keyword "sustainable manufacturing" has become very popular. Sustainable manufacturing considers environmental, economic, and social aspects. The aim of the sustainable manufacturing is to develop technologies for transforming materials with objectives of reducing emission of greenhouse gases, avoiding use of nonrenewable or toxic materials, and avoiding generation of waste. In a keynote paper, Westkamper et al. (2000) have listed the following five critical factors that the world is confronting:

1. A rising consumption of natural resources
2. The dramatic increase in world population
3. Environmental impacts, i.e., limited natural resources (energy, materials)
4. Global communication networks based on standards
5. An unstoppable worldwide globalization

To overcome these factors, there is a need to switch over to sustainable manufacturing. The term sustainability is often used to cover environmental issues.

As a result of the UN General Assembly resolutions, the World Commission on Environment and Development was set up. The commission published a report entitled "Our Common Future," which is commonly known as Brundtland report. It provided the following definition of sustainability:

> Sustainable development meets the needs of the present without compromising the abilities of future generations to meet their own needs.

Thus, sustainability is the ability to maintain the desired living conditions for all times to go.

There have been several suggestions for ensuring sustainable development. Population growth needs to be controlled. Natural resources have to be used judiciously. Gaseous, liquid, and solid wastes have to be minimized. In fact, "by placing monetary values on natural resources, environmental quality, and degradation, it is possible to incorporate these factors into the economic analysis of projects, and perhaps on a wider scale, into the national system of accounts" (Toakley and Aroni 1998). It is apparent that environmental considerations are the major part of a sustainable manufacturing system.

1.4   What is Green Manufacturing?

Of late, there has been an increasing concern about environmental consideration in manufacturing (Sheng and Srinivasan 1995). Environmentally conscious manufacturing (ECM) is concerned with developing methods for manufacturing new product from conceptual design to final delivery and ultimately to the end-of-life disposal such that environmental standards and requirements are satisfied (Gungor and Gupta 1999). The manufacturing industry is one of the main causes of environmental pollution. How to minimize the environmental impact of the manufacturing industry is an important concern for all manufacturers. During this critical time, an advanced manufacturing mode called green manufacturing (GM) has become popular as a sustainable development strategy in industrial processes and products. GM is a modern manufacturing strategy, which is crucial for the twenty-first century manufacturing industries, integrating all the issues of manufacturing with its ultimate goal of reducing and minimizing environmental impact and resource consumption during a product life cycle. GM is a method for manufacturing that minimizes waste and pollution. These goals are often achieved through product and process design.

In most of the factories, there is no comprehensive evaluation tool for green attribute of manufacturing process. Krishnan and Sheng (2000) presented an automatic process-planning agent for CNC machining for minimal environmental impact. The process-planning system can accept web-based designs and offers feedback to the designer over the Internet. Tan et al. (2002) carried out research work on green manufacturing for several years, and have proposed a decision-making framework model for green manufacturing. They considered several decision-making objective factors such as time (T), quality (Q), cost (C), environmental impact (E), and resource consumption (R) in their model. Yan et al. (2007) presented a process-planning support system for green manufacturing (GMPPSS) to deal with the problems in optimization of environment-favorable process planning. The objective of the GMPPSS is to evaluate the green attributes of the process planning from the single process elemental level to the entire process project level, which include raw material consumption, secondary material consumption, energy consumption, and environmental impacts of the manufacturing process.

Environmental consciousness in society has been increasing day by day, which reflects in manufacturing with numerous opportunities to improve the environmental performance of manufacturing. In his book, Kutz (2007) defines the ECM process as the production of products using processes and techniques selected to be both economically viable and having the least impact on the environment. Three basic objectives have been mentioned for consideration of ECM—reduction of waste, reduction of hazardous materials and processes, and reduction of energy. Ilgin and Gupta (2010) reviewed more than 540 papers on the work of ECM and product recovery for the last 1 decade to search about the avenues for future research. One of their conclusions is as follows. "Product Design research mainly focuses on multi criteria techniques which allow for the simultaneous consideration of environmental, economic, consumer and material requirements. However, the environmental impact of production processes is ignored in most of these studies. Thus, there is a need for environmentally conscious product design methodologies that integrate design of products and processes."

1.5 Environmentally Conscious Machining

Generally, in different machining processes, offensive pollutants and by-products are generated which should be treated properly to protect the environment. Many a times, cutting fluid is used in machining to enhance the tool life and/or improve the surface integrity. The fluids that are used to lubricate in machining contain environmentally harmful or potentially damaging chemical constituents. Prolonged exposure to coolants during machining may lead to respiratory irritation, asthma, pneumonia, dermatitis, cancer, etc. (Burge 2006). Most of the cutting fluids in machining are petroleum-based oils. The disposal of petroleum-based oils is problematic. They cause surface water and ground water contamination, air pollution, and soil pollution. The effect is transmitted to agricultural produce and the food consumed by us. Skin contact with these fluids may cause irritation and allergy. During the storage of water-soluble cutting fluids, microbial toxins are generated by bacteria and fungi present in the surroundings. Nowadays research is going on to use vegetable-based oils as cutting fluids (Shashidhara and Jayaram 2010).

Apart from eliminating or minimizing the dangerous cutting fluids in machining, minimization of energy consumption should be another objective of environmentally conscious machining. This is because energy consumption is linked with environmental pollution, since almost all energy-generating methods pollute the environment in some way. This calls for the optimization of machining processes. Thus, optimization of machining processes is a must for an environmentally friendly machining process.

In many machining processes, the cutting tools are not utilized properly and are thrown away before their life is over. Moreover, a non-optimized machining process causes more tool wear and consequently increased consumption of cutting tools. Increased consumption of tool increases the expenditure due to procurement of tools as well as causes problems related to disposal of the tool.

In this monograph on environmentally friendly machining, all the above-mentioned aspects are considered. A proper strategy of environmentally friendly machining needs to minimize energy, cutting fluid, and cutting tools. The strategy can be made effective with the help of modern modeling and optimization tools.

1.6 Conclusion

In this chapter, the concept of environmentally friendly machining has been introduced. A brief description of green engineering, sustainable manufacturing, green manufacturing, and environmentally conscious machining has been provided. The word "environmentally friendly machining" is usually considered synonymous with "environmentally conscious machining." The rest of the monograph discusses the methods to achieve the goals of environmentally friendly machining.

References

Abraham MA, Nguyen N (2003) Green engineering: defining the principles, Environmental Progress, 22, pp233–236. Extracts used in the chapter available at: http://www.epa.gov/opptintr/greenengineering/pubs/whats_ge.html. Accessed on 8 May 2011

Anastas PT, Zimmerman JB (2003) Design through the twelve principles of green engineering. Environ Sci Technol 37:94A–101A

Burge H (2006) Machining coolants, The Environmental Reporter, technical newsletter, EMLab 4. http://www.emlab.com/s/sampling/env-report-10-2006.html. Accessed on 8 May 2011

Gungor A, Gupta SM (1999) Issues in environmentally conscious manufacturing and product recovery: a survey. Comput Ind Eng 36:811–853

Ilgin MA, Gupta SM (2010) Environmentally conscious manufacturing and product recovery (ECMPR): a review of the state of the art. J Environ Manage 91:563–591

Krishnan N, Sheng PS (2000) Environmental versus conventional planning for machined components. CIRP Ann Manuf Technol 49:363–366

Kutz M (2007) Environmentally conscious manufacturing. Wiley, Hoboken, NJ

Shashidhara YM, Jayaram SR (2010) Vegetable oils as a potential cutting fluid—an evolution. Tribol Int 43:1073–1081

Sheng P, Srinivasan M (1995) Multi-objective process planning in environmentally conscious manufacturing: a feature-based approach. Ann CIRP 44:433–437

Tan XC, Liu F, Cao HJ, Zhang H (2002) A decision-making framework model of cutting fluid selection for green manufacturing and a case study. J Mater Process Technol 129:467–470

Toakley AR, Aroni S (1998) Forum the challenge of sustainable development and the role of universities. Higher Educ Pol 11:331–346

U.S. Environmental Protection Agency, Green Engineering. http://www.epa.gov/opptintr/greenengineering/pubs/whats_ge.html. Accessed on 8 May 2011

Westkamper E, Alting L, Arndt G (2000) Life cycle management and assessment: approaches and visions towards sustainable manufacturing (keynote paper). CIRP Ann Manuf Technol 49:501–526

World Commission on Environment and Development (1987) Our common future. Oxford University Press, Oxford

Yan HE, Fei L, Huajun C, Hua Z (2007) Number of electrons. Front Mech Eng China 2:104–109

Chapter 2
Machining with Minimal Cutting Fluid

Abstract The purpose of cutting fluid in a machining operation is to cool the workpiece, reduce friction, and wash away the chips. The cutting fluid contributes significantly toward machining cost and also possesses environmental threats. In the past, there have been some attempts to minimize the amount of cutting fluid in machining. This chapter reviews some prominent ways to minimize the application of cutting fluid and their impact on the machining performance.

Keywords Cutting fluids • Disposal of cutting fluids • Internal coolant supply • Minimal quantity lubrication • Minimum quantity cooling lubrication • Mist lubrication • Nanofluids • Near dry grinding • Straight cutting oils • Water-mix fluids

2.1 Introduction

Cutting fluids are employed in machining to reduce friction, cool the workpiece, and wash away the chips. With the application of cutting fluid, the tool wear reduces and machined surface quality improves. Often the cutting fluids also protect the machined surface from corrosion. They also minimize the cutting forces thus saving the energy. These advantages of using cutting fluids in machining are accompanied by a number of drawbacks. Sometimes the cutting fluid costs are more than twice the tool-related costs (Astakhov 2008). Most of the cutting fluids possess the health hazard to the operator. Disposal of the used cutting fluid is also a major challenge.

In the recent past, there has been a general liking for dry machining (Sreejith and Ngoi 2000). On the other hand, several researchers started exploring the application of minimal cutting fluid. In this chapter, a review of the application of minimal cutting fluid in machining is presented.

U.S. Dixit et al., *Environmentally Friendly Machining*, SpringerBriefs
in Applied Sciences and Technology, DOI 10.1007/978-1-4614-2308-9_2,
© Springer Science+Business Media, LLC 2012

2.2   Major Concerns in Using Cutting Fluids

There are mainly two types of cutting fluids used in machining (1) neat oils or straight cutting oils (2) water-mix fluids. Neat oils are based on mineral oils and used for the metal cutting without further dilution. They are generally blends of mineral oils and other additives. The most commonly used additives are fatty materials, chlorinated paraffin, sulfurized oils, and free sulfur. Sometimes organic phosphorous compounds are also used as additives. Extreme pressure additives containing chlorine, sulfur, or phosphorous react in the tool–chip interface producing metallic chlorides, phosphates, and sulfides, thus protecting the cutting edge (Trent 1984). Neat oils provide very good lubrication but poor cooling. Water-mix fluids are of three types (a) emulsifiable oils (b) pure synthetic fluids (c) semisynthetic fluids. Emulsifiable oils form an emulsion when mixed with water. They are used in a diluted form with concentration of 3–10%. The concentrate consists of a base mineral oil and emulsifiers. These oils produce good lubrication and cooling. Pure synthetic fluids contain no petroleum or mineral oil base and are formulated from alkaline inorganic and organic compounds with additives for corrosion inhibition. They are used in a diluted form with concentration of 3–10%. They provide very good cooling performance. Semisynthetic fluids are the mixture of emulsifiable oils and pure synthetic fluids. Their characteristics are the mix of the characteristics of emulsifiable and pure synthetic oils.

Cutting fluids often pose hazard to man, machine, and material. For example, a cutting fluid with fatty material reacts with the zinc and produces zinc soap. Hence, the use of galvanized tanks, pipes, and fittings should be avoided with it. Fatty oil based fluids readily oxidize, particularly in the presence of a catalyst like copper. Thus, during the machining of copper, fat is converted to organic acid which reacts with exposed copper surface to produce green color copper soaps. Presence of chlorine also poses health hazard. Sulfur also reacts with many metals to make sulfides.

Water-mix fluids cause staining and corrosion. They also produce microorganisms. All water-mix cutting fluids are alkaline for inhibiting the corrosion. It also helps to control the growth of microorganisms. However, excessive alkalinity causes irritation to human skin. It also causes corrosion problems in aluminum and zinc. As the magnesium is very reactive with water, it should not be machined with water-mix fluid. Synthetic fluids usually contain triethanolamine which reacts with copper. They are also not suitable for machining of aluminum. Hope (1977) has reviewed staining and corrosion tendencies of cutting fluids.

A number of occupational diseases of operators are due to skin contact with cutting fluids. Direct skin contact can cause an allergic reaction or dermatitis. The petroleum products that are basis for the majority of the fluids are suspected carcinogens. It was noted that machinists exhibited a higher rate of upper respiratory tract cancer than other workers. Applying the cutting fluid in the form of oil mist also poses serious health hazards. The contact of mist with eye may cause irritation and the mist may affect adversely to asthma patients. It may also cause long time breathing disorders. According to the Occupational Safety and Health

Administration (OSHA), the permissible exposure level of mist within the plant is 5 mg/m^3, which may be reduced to 0.5 mg/m^3. Another problem with the use of cutting fluids is that in many cases, the harmful effects of the cutting fluids are not known due to lack of studies (Bennett 1983). Some studies have indicated that the respiratory exposure to ozone or nitrogen oxide in combination with exposure to oil mists increases the toxic effects of the oxidants. The toxicity of formaldehyde vapors increases in the presence of nontoxic aerosols (mists) of mineral oils or glycols.

The disposal of cutting fluids is also a big problem. The waste cutting fluids can pollute surface and groundwater. They can cause soil contamination, affect agriculture produce, and can lead to food contamination. Thus, ideally, cutting fluids should not be used at all. If it is not possible, then their use should be minimized. One alternative is to develop completely safe cutting fluids, but they may not be competitive due to economic consideration.

2.3 Minimum Quantity Lubrication Systems

The conventional system of applying the coolant is flood coolant system, in which a large quantity of coolant is continuously impinged on the rake face of the tool. This system is very inefficient. First of all, a large quantity of the cutting fluid is required. Second, the cutting fluid is not able to reach the cutting zone due to obstruction from chips.

A better method is the application of mist lubrication, in which a mixture of air and cutting called aerosol is produced and supplied in the cutting zone with a high pressure. The system uses an atomizer. The atomizer is an ejector where the compressed air is used to atomize the cutting oil (Fig. 2.1). Oil is then conveyed by the air in a low-pressure distribution system to the machining zone. As the compressed air flows through the venturi path, the narrow throat around the discharge nozzle creates a venturi effect in the mixing chamber, i.e., a zone where the static pressure is below the atmospheric pressure (often referred to as a partial vacuum). This partial vacuum draws the oil up from the oil reservoir where the oil is maintained under a constant hydraulic head. The air rushing through the mixing chamber atomizes the oil stream into an aerosol of micron-sized particles. When the aerosol impinges through the jet, it produces a spray of gaseous suspension called mist in the machining zone which works as cooling as well as lubricating medium. However, mist also poses a health hazard.

Instead of applying the cutting fluid from an external nozzle, channels can be made in the tool for supply of cutting fluid to the high temperature zone. Figure 2.2 is a schematic representation of such type of tool, in which the high pressure coolant is forced through a hole to reach the cutting face of the tool. This type of arrangement was used about 2 decades earlier by Wertheim et al. (1992). In their arrangement, the cutting fluid is able to reach the cutting zone more effectively than through external application. In the conventional flood coolant system, the heat

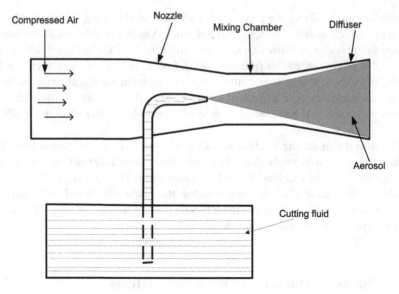

Fig. 2.1 Schematic of an atomizer

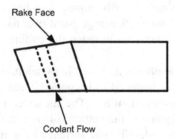

Fig. 2.2 A cutting tool with an internal coolant supply

causes the evaporation of the coolant before it reaches the critical area. Wertheim et al. (1992) used a high pressure system. The pressure was increased up to 25 bar. This system reduced tool wear and improved the tool life. In the grooving in alloy steel by TiC + TiCN + TiN coated carbide tool, it was observed that tool could make 40 grooves when the flushing pressure was 1 bar. With 5-bar pressure, the tool life was 75 grooves and with 25-bar pressure, a total of 160 grooves were produced before the tool failed. Similar phenomenon was observed when grooving a high temperature alloy, Inconel 718. When using a cutting speed of 30 m/min at a feed of 0.16 mm/rev, a tool life of only 3 min was reached with conventional flushing. Using the internal flushing at 16 bar under the same machining conditions, a total tool life of over 14 min was achieved. Compared to conventional system, the requirement of the coolant got drastically reduced. Kovacevic et al. (1995) studied the performance of a face milling process, in which a high pressure water jet was

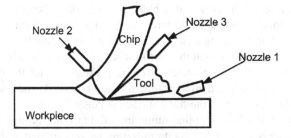

Fig. 2.3 MQL application with three jets

delivered into tool–chip interface through a hole in the tool rake face. Senthil Kumar et al. (2002) applied a high pressure (17 bar) cooling system in high speed milling, in which the cutting fluid was supplied though the spindle.

Weinert et al. (2004) have presented an excellent review of dry machining and minimum quantity lubrication (MQL). When the primary objective is to carry out lubrication, the system is MQL. When both cooling and lubrication are needed, it is called Minimum Quantity Cooling Lubrication (MQCL). In MQCL operations, the media used is generally straight oil, but some applications have used an emulsion or water. The cutting oil can be sent with air in the form of aerosol or without air. In MQL or MQCL system, the normal consumption of cutting fluid medium is 5–50 ml/min. The supply of cutting fluid can be external (through nozzles) or internal (through a channel) in tool. There can be a single channel system or double channel system, in which the air and oil are fed separately.

The fluid used in MQL or MQCL system should be biodegradable and stable. As the consumption of the oil is very less, the fluid should remain stable for a longer period of time. Vegetable oils and synthetic esters have been used as cutting fluids in MQL applications (Khan and Dhar 2006; Wakabayashi et al. 2003). A synthetic ester has a high boiling temperature and flash point and a low viscosity and thus leaves a thin film of oil on the workpiece which serves to resist corrosion. Synthetic esters are biodegradable also.

Method of applying the cutting fluid has a great effect on machining performance in an MQL system. In an orthogonal machining, cutting fluid can be injected at three places through different nozzles as shown in Fig. 2.3. Cutting fluid injected through Nozzle 1 reduces the friction between tool and workpiece and helps in reducing flank wear. The injection of fluid at Nozzle 2 helps in the curling of chips because of Rebinder effect and cooling. Here, some heat from primary shear zone is taken away. The injection through Nozzle 3 helps in taking the heat away from secondary shear zone on the rake face.

Varadarajan et al. (2002) supplied the specially formulated cutting fluid in the form of thin pulsed jet in the hard turning process. A fuel pump of diesel engine was used for injection. The system can deliver cutting fluids through six outlets simultaneously, but the study was conducted with one outlet only. The nozzle position approximately corresponded to Nozzle 1 in Fig. 2.3. The typical rate of discharge was 2 ml/min. The jet velocity is of the order of 100 m/s at a pressure of 200 bar.

The fluid was injected in pulses at a pulse rate of 600 pulses/min. The system provided very good performance in terms of cutting forces, cutting temperature, tool life, surface finish, cutting ratio, and tool–chip contact length.

Attanasio et al. (2006) studied the performance of MQL turning by injecting the lubricants on rake and flank separately. Referring Fig. 2.3, once the lubricant was injected through Nozzle 1 and another time through Nozzle 3. The conclusion was that injecting the lubricant on the flank surface is better.

Ram Kumar et al. (2008) applied minimum cutting fluid through two jets in a hard turning process. One high velocity pulsating jet was applied at the tool–work interface and other was applied on the top surface of the chip as shown (corresponding to Nozzle 1 and 3 in Fig. 2.3). This causes the curling of the chip due to difference in the top and bottom surface temperatures. Thus, the chip-tool contact length is reduced, helping to reduce the cutting force and temperature and thus improving the tool life. In this system, the pressure of the cutting fluid was kept at 1.2 bar and the amount of cutting fluid was 5–10 ml/min. The pump was operated at 300–600 pulses/min. The system provided reduced surface finish, tool wear, cutting force, cutting temperature, and tool–chip contact length.

2.4 Minimum Quantity Lubrication with Nanofluids

Nanofluids are the fluids with a colloidal dispersion of nanometer-sizes particles of metals, oxides, carbides, nitrides, or nanotubes. Typically, a nanofluid may contain carbon nanotube (CNT), TiO_2, Al_2O_3, MoS_2, and diamond. Size of the nanoparticles is between 1 and 100 nm. Nanofluids show enhanced thermal conductivity and heat transfer coefficient. With the addition of nanoparticles, the thermal conductivity of the fluids can enhance by several hundred percents. This is mainly due to more surface-to-volume ratio of nanoparticles.

Recently, nanofluids have been used with MQL systems. The nanofluid is supplied to the machining area in the form of mist mixed with highly pressurized compressed air. Nanofluids have been containing MoS_2, diamond, and Al_2O_3 in grinding and milling. Nam et al. (2011) applied nanofluid containing 30-nm size diamond particles with the base fluids of paraffin and vegetable oils in microdrilling of aluminum 6061 workpiece. The performance of nanofluid MQL was compared with compressed air lubrication and pure MQL. The nanodiamond concentration of 1% and 2% by volume was considered for study. The addition of nanodiamond particles improved lubrication and cooling effects with their enhanced penetration and entrapment at the drilling interface. It is reported that nanoparticles have ball/rolling bearing effect and enhance tribological and wear characteristics significantly. As a result, the magnitude of torques and thrust forces were significantly reduced. The authors observed that paraffin oil based nanofluid MQL was more effective than the vegetable oil based one. In the case of the paraffin oil, the 1 vol% of nanodiamond particles was more effective than 2 vol% of particles. On the other hand, in the case of the vegetable oils, the nanofluid with 2 vol% was found better.

Authors attributed this to difference in the physical and chemical properties of two base fluids. In particular, the dynamic viscosity of vegetable oils is about 2–3 times higher than that of paraffin. Therefore, more nanoparticles could be needed for getting evenly spread in the drilling area. In the case of paraffin oils, 2 vol% may cause some nanodiamond particles to get clogged.

2.5 Minimum Quantity Lubrication: A Comparison with Other Systems

There is enough literature to show that MQL system provides better performance than dry machining. In many cases, it provides better performance than conventional flood coolant system. A brief representative review is provided in this section. When machining aluminum alloys, Kelly and Cotterell (2002) observed that as cutting speed and feed rate are increased, the use of a fluid mist outperformed the conventional flood coolant method, however, at lower cutting speed flood coolant system was superior. Braga et al. (2002) used a spray mist while drilling aluminum alloy and observed that surface finish and tool life was almost same in mist lubrication and flood coolant.

Mendes et al. (2006) studied the performance of drilling of AA 1050-O aluminum with TiAlN coated carbide drills and applied cutting fluid as mist. The cutting fluid flow rate was varied between 20 and 100 ml/h. It was observed that using the highest cutting fluid flow rate (100 ml/h) resulted in lower feed forces only at higher cutting speeds and feed rates. Power consumption and specific cutting pressure increased with cutting fluid flow rate and surface roughness was unaffected. This work shows that unnecessarily higher fluid flow rate is not useful. Davim et al. (2006) studied drilling of aluminum (AA1050) under dry, MQL and flood-lubricated conditions and concluded that with proper selection of cutting parameters, it is possible to obtain machining performance similar to flood-lubricated conditions by using MQL. Davim et al. (2007) made similar conclusion in turning of brasses using MQL.

In the turning of 6061 aluminum alloy with MQL, dry and flood lubricant conditions using diamond-coated carbide tools, Sreejith (2008) observed the superiority of MQL with 50 ml/h and 100 ml/h cutting fluid consumption. The tool wear was almost same as in the flood coolant system. The main cutting force was the lowest in the flood coolant system and the highest in dry machining. The surface roughness with 100 ml/h MQL was much lower than that obtained in dry machining. It was only slightly greater than the surface roughness obtained in flood coolant system.

Tawakoli et al. (2009) have investigated an MQL grinding or near dry grinding (NDG) system. In this system, an air–oil mixture called an aerosol is fed into the wheel-work zone. Compared to dry grinding, MQL grinding substantially enhances cutting performance in terms of increasing wheel life and improving the quality of the ground part. In the grinding of 100Cr6 hardened steel by Al_2O_3 grinding wheel, the surface roughness of ground part was lower than that in flood coolant system. However, in MQL grinding of 42CrMo4 soft steel, the surface roughness was higher than that in flood coolant system. In MQL grinding the cutting forces were lower than the dry and flood coolant systems. The wheel life was the best in MQL systems.

Alberdi et al. (2011) optimized the nozzle design in MQL grinding with the help of computational fluid dynamics. The optimized nozzle provides a more efficient coolant jet. The authors also proposed a technique based on the combination of MQL and low temperature CO_2 to assist grinding. The significant improvement in performance was obtained compared to other systems. The authors recommended that with the proposed system, the grinding wheels of higher porosity should be used for getting the best results.

2.6 Conclusion

From the discussion presented in this chapter, it is apparent that MQL systems possess many advantages over flood coolant system. However, they also require some modification of machine tools for obtaining the best performance out of them. When the flood coolant system is not present, the machine tools should be equipped with a chip removal system. There is also a requirement of fire and explosion system in the machining of light metal alloys like magnesium. There is additional cost involved in the equipment for MQL. A cost-benefit analysis is required before implementing MQL system.

References

Alberdi R, Sanchez JA, Pombo I, Ortega N, Izquierdo B, Plaza S, Barrenetxea D (2011) Strategies for optimal use of fluids in grinding. Int J Mach Tools Manuf 51:491–499
Astakhov VP (2008) Ecological machining: near-dry machining. In: Davim JP (ed) Machining: fundamentals and recent advances. Springer, London
Attanasio A, Gelfi M, Giardini C, Remino C (2006) Minimal quantity lubrication in turning: effect on tool wear. Wear 260:333–338
Bennett EO (1983) Water based cutting fluids and human health. Tribol Int 16:133–136
Braga DU, Diniz AE, Miranda GWA, Coppini NL (2002) Using a minimum quantity of lubricant (MQL) and a diamond coated tool in the drilling of aluminium-silicon alloys. J Mater Process Technol 122:127–138
Davim JP, Sreejith PS, Gomes R, Peixoto C (2006) Experimental studies on drilling of aluminium (AA1050) under dry, minimum quantity of lubricant, and flood-lubricated conditions. Proc IMechE B J Eng Manuf 220:1605–1611
Davim JP, Sreejith PS, Silva J (2007) Turning of brasses using minimum quantity of lubricant (MQL) and flooded lubricant conditions. Mater Manuf Process 22:45–50

Hope DA (1977) Cutting fluids—pet or pest?: a review of staining and corrosion tendencies and effects on machine tool paints and seals. Tribol Int 10:23–27

Kelly JF, Cotterell MG (2002) Minimal lubrication machining of aluminium alloys. J Mater Process Technol 120:327–334

Khan MMA, Dhar NR (2006) Performance evaluation of minimum quantity lubrication by vegetable oil in terms of cutting force, cutting zone temperature, tool wear, job dimension and surface finish in turning AISI-1060 steel. J Zhejiang Univ Sci A 7:1790–1799

Kovacevic R, Cherukuthota C, Mazurkiewicz M (1995) High pressure waterjet cooling/lubrication to improve machining efficiency in milling. Int J Mach Tools Manuf 35:1459–1473

Mendes OC, Avila RF, Abrao AM, Reis P, Davim JP (2006) The performance of cutting fluids when machining aluminium alloys. Ind Lubric Tribol 58:260–268

Nam JS, Lee PH, Lee SW (2011) Experimental characterization of micro-drilling process using nanofluid minimum quantity lubrication. Int J Mach Tools Manuf 51:649–652

Ram Kumar P, Leo Dev Wins K, Robinson Gnanadurai R, Varadarajan AS (2008) Investigations on hard turning with minimal multiple jet of cutting fluid. Proceedings of the international conference on frontiers in design and manufacturing engineering, Karunya University, Coimbatore, 1-2 February 2008

Senthil Kumar A, Rahman M, Ng SL (2002) Effect of high-pressure coolant on machining performance. Int J Adv Manuf Technol 20:83–91

Sreejith PS (2008) Machining of 6061 aluminium alloy with MQL, dry and flooded lubricant conditions. Mater Lett 62:276–278

Sreejith PS, Ngoi BKA (2000) Dry machining: machining of the future. J Mater Process Technol 101:287–291

Tawakoli T, Hadad MJ, Sadeghi MH, Daneshi A, Stockert S, Rasifard A (2009) An experimental investigation of the effects of workpiece and grinding parameters on minimum quantity lubrication—MQL grinding. Int J Mach Tools Manuf 49:924–932

Trent EM (1984) Metal cutting, 2nd edn. Butterworths & Co., London

Varadarajan AS, Philip PK, Ramamoorthy B (2002) Investigations on hard turning with minimal cutting fluid application (HTMF) and its comparison with dry and wet turning. Int J Mach Tools Manuf 42:193–200

Wakabayashi T, Inasaki I, Suda S, Yokota H (2003) Tribological characteristics and cutting performance of lubricant esters for semi-dry machining. CIRP Ann Manuf Technol 52:61–64

Weinert K, Inasaki I, Sutherland JW, Wakabayashi T (2004) Dry machining and minimum quantity lubrication. CIRP Ann Manuf Technol 53(2004):511–537

Wertheim R, Rotberg J, Ber A (1992) Influence of high-pressure flushing through the rake face of the cutting tool. CIRP Ann Manuf Technol 41:101–106

Chapter 3
Dry Machining

Abstract It is possible to eliminate the cutting fluids totally in machining. Dry machining refers to machining without using cutting fluids. In this chapter, advantages and limitation of dry machining are described. Technological requirement for dry machining are elaborated.

Keywords Chip handling • Cryogenic cooling • Dry machining • Heat pipe • Lubrication • Minimal quantity lubrication • Near dry machining • Peltier effect • Self-lubricating tools • Under cooling system

3.1 Introduction

Environmental concerns call for the elimination of cutting fluid in metal cutting practice. In the recent past, a lot of interest is being taken in the machining without using the cutting fluid or using the minimal amount of cutting fluid. When no cutting fluid is used during machining, it is called dry machining. When a minimal amount of cutting fluid is used during machining, it is called near dry machining or minimum quantity lubrication (MQL). Near dry machining has been discussed in Chap. 2.

The MQL system uses either aerosol (a colloidal suspension of cutting fluids droplets in air) or directly the metered quantity of cutting fluid at the cutting zone. Usually, the cutting fluid gets vaporized due to high temperature at the cutting zone. Ultimately, the cutting fluid mixes with atmospheric air and is inhaled by the operator. Occupational Safety and Health Administration (OSHA) has prescribed 5.0 mg/m^3 as the permissible exposure limit (PEL) of the cutting fluid in air. The United Auto Workers petitioned the OSHA to lower the PEL for metal working fluids from 5.0 mg/m^3 to 0.5 mg/m^3. In response, OSHA established the Metal Working Fluid Standards Advisory Committee (MWFSAC) in 1997 to develop standards or guidelines related to metal working fluids. Its objective was to investigate the need for and/or recommend, if appropriate, a standard, guideline, or other

U.S. Dixit et al., *Environmentally Friendly Machining*, SpringerBriefs
in Applied Sciences and Technology, DOI 10.1007/978-1-4614-2308-9_3,
© Springer Science+Business Media, LLC 2012

appropriate response to the health effects, if any, that result in material impairment to workers occupationally exposed to metal working fluids. In its final report in 1999, MWFSAC recommended that the exposure limit be 0.5 mg/m^3 and that medical surveillance, exposure monitoring, system management, workplace monitoring, and employee training are necessary to monitor worker exposure to metal working fluids.

Dry machining is the best solution from an environmental point of view. The dry machining has the advantage of non-pollution of the water and air, no problem in disposal of the cutting fluid, and no danger to health of the operator. However, with dry machining one has to adopt the measures suitable to compensate the primary functions of the cooling lubricant.

3.2 Advantages of Dry Machining

The following are the main advantages of the dry machining:

1. Dry machining does not cause the pollution of atmosphere or water. In wet machining (using a large amount of coolant) or MQL machining, environment gets polluted due to dissociation of cutting fluid. During disposal of the cutting fluid, water gets polluted and soil gets contaminated. These problems are absent in dry machining.
2. During metal cutting operation, a lot of solid waste in the form of debris is generated. This is called swarf and comprises chips, metallic dust, and small metal pieces. In the sustainable manufacturing practice, it is advisable to recycle the swarf. During machining with cutting fluid, the swarf contains residue of cutting fluid, which needs to be removed by costly chemical treatment. In dry machining, there is no residue on the swarf. Thus, recycling the swarf is easier. In fact, dry swarf can be sold at higher price than the wet swarf.
3. There is no danger to the health of operator in dry machining. It is non-injurious to skin and is allergy free.
4. It is estimated that about 15% of the machining cost can be attributed to the use of coolant, which is about 3–4 times the cutting tools cost (Landgraf 2004). In addition to the coolant itself, the coolant related cost includes disposal, storage, maintenance and labor cost components. Dry machining will save coolant related cost.
5. Sometimes, the dry machining requires less cutting force compared to machining with coolant. This is specially, true for high speed machining, in which high temperature generated at cutting zone lowers the flow stress of material.
6. In many cases, for example in interrupted cutting, dry cutting improves tool life. In machining with cutting fluid, the tool is subjected to thermal shock, which lowers tool life. The process such as milling, in which the tool does not cut continuously are better suited for dry machining.

3.3 Technologies in Dry Machining

The cutting fluid has the three main functions—removal of heat, lubrication, and washing away of the chips. Among these, the chip removal is the most crucial function in drilling. The main problem with dry drilling is the removal of chips from the drilled hole. In the following subsections, alternative methods for achieving the functions of cutting fluid are discussed.

3.3.1 Removal of Heat

There are a number of ways of removing heat from the tool in dry machining. One way is to cool the base of the tool, which is called under cooling (Ber and Goldblatt 1989). In under cooling system, the coolant flows through channels located under the insert as shown in Fig. 3.1. In order to remove maximum heat, a thin copper foil is in direct contact with the cutting insert in its upper side, while, its opposite side, is in contact with the coolant. The copper foil has two protrusions, bent downwards, in order to shift the fluid away from the cutting zone and the workpiece material. Thus, the effective cooling area gets increased.

Another method is internal cooling with an evaporation system, where a volatile liquid is introduced into the tool holder that evaporates in contact with the interior system of the insert. Sanchez et al. (2011) used R-123, a hydrochloroflurocarbon fluid as the coolant. It has boiling temperature of 28°C. The liquid is evaporated after coming in contact with the internal surface of cutting tool and reaches the condenser. In the condenser, it gets converted into liquid form and reaches a reservoir, from where it is pumped to tool holder. It is a closed circuit system.

As per Montreal Protocol on substances that deplete the ozone layer, the hydrochloroflurocarbon fluids should not be used beyond 2015. Thus, other suitable fluids have to be used in the under cooling system, the basic principle remaining same. Zhao et al. (2002) has developed a mathematical model for tool wear prediction in the presence of internal cooling. It is found that internal cooling in the carbide cutting tool can increase the tool life up to 15% compared with the cutting tool without internal cooling.

There have been some attempts to use heat pipe for internal cooling of the cutting tool. A heat pipe is a device that uses thermal conductivity and phase transition to transfer the heat from one place to other place. It consists of an enclosed pipe containing a liquid and its vapor. At the hot interface that acts as evaporator, the liquid gets converted to vapor by absorbing heat from the hot surface. At the cold interface that acts as condenser, the vapors condense. The liquid then returns to hot interface through either capillary action or gravity action, where it is evaporated again. This cycle keeps repeating and heat removal takes place.

Jen et al. (2002) investigated the feasibility of using heat pipe in the drilling process by conducting theoretical and experimental simulations. The schematic of the proposed heat pipe is shown in Fig. 3.2. The components of a heat pipe are a

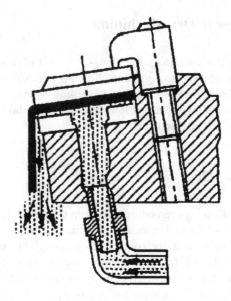

Fig. 3.1 An under cooling system. With permission from Ber and Goldblatt (1989). Copyright (1989) Elsevier

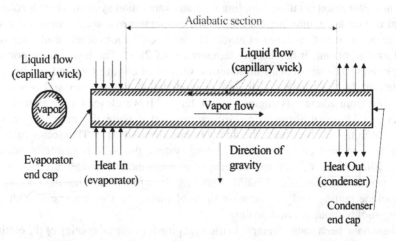

Fig. 3.2 Schematic of proposed heat pipe. With permission from Jen et al. (2002). Copyright (2002) Elsevier

sealed container (pipe wall and end caps), a wick structure, and a small amount of working fluid in equilibrium with its own vapor. The heat pipe can be divided into three sections: evaporator section, adiabatic (transport) section, and condenser section. The external heat load on the evaporator section causes the working fluid to vaporize. The resulting vapor pressure drives the vapor through the adiabatic section to the condenser section, where the vapor condenses, releasing its latent heat of vaporization to the provided heat sink. The condensed working fluid is then

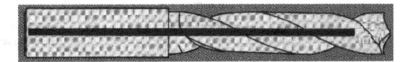

Fig. 3.3 Schematic of drill with internally embedded heat pipe. With permission from Jen et al. (2002). Copyright (2002) Elsevier

pumped back by capillary pressure generated by the meniscus in the wick structure. Transport of heat can be continuous as long as there is enough heat input to the evaporator section such that sufficient capillary pressure is generated to drive the condensed liquid back to the evaporator. Figure 3.3 shows schematically how this type of heat pipe can be embedded in a drill. Numerical studies and initial experiments showed that the use of a heat pipe inside the drill reduces the temperature field significantly.

Cryogenic cooling is another method of removing heat in dry machining. Cryogenics is the study of the production of very low temperature (below $-150°C$) and the behavior of materials at those temperatures. A number of gases in liquid form can be used to achieve cryogenic temperatures. Among them liquid nitrogen is the most widely used one. Nitrogen melts at $-210°C$ and boils at $-196°C$. It is the most abundant gas in atmosphere, about 78% by volume of the atmosphere. It is a colorless, odorless, tasteless, and nontoxic gas. It does not react with most metals. Because of these qualities, nitrogen is a preferred choice as coolant. Yildiz and Nalbant (2008) have reviewed the application of cryogenic cooling in machining processes. An early application of cryogenic machining is due to Hollis (1961), who introduced liquid CO_2 to the base of the carbide tip through a calibrated capillary tube carried in the tool shank to provide a low ambient temperature and an increased temperature gradient through the cross section of the tip material. This reduced fracture and wear of the tool. Paul and Chattopadhyay (1995) applied liquid nitrogen jet in the grinding process. This reduced the grinding forces, specific cutting energy, grinding temperature, and residual stresses on workpiece. The performance of cryogenic cooling was compared with dry grinding and grinding with flood coolant, and it was found superior to both of them.

Wang and Rajukar (2000) carried out cryogenic machining of hard-to-machine materials. Authors presented a technique for machining of advanced ceramics with liquid nitrogen cooled polycrystalline cubic boron nitride (CBN) and cemented carbide tool. It was found that with liquid nitrogen cooling, the temperature in the cutting zone is reduced substantially. This significantly reduced the tool wear. The surface roughness also got reduced. Paul et al. (2001) studied the cryogenic machining of AISI 1060 steel using two different types of carbide inserts. Two liquid nitrogen jets were employed one at the flank surface and other at rake surface of the cutting tool. The authors have not mentioned about the flow rate and pressure of liquid nitrogen. A significant improvement in the surface finish of the workpiece and tool life was obtained. Tool life got increased by a factor of 2.5. More recently, Kalyan Kumar and Choudhury (2008) applied cryogenic cooling to high speed

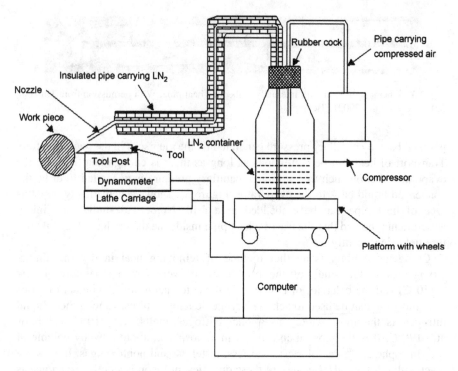

Fig. 3.4 Schematic of the experimental set up. With permission from Kalyan Kumar and Choudhury (2008). Copyright (2008) Elsevier

machining of stainless steel using a standard insert. A schematic of their experimental setup is shown in Fig. 3.4. The liquid nitrogen jet was supplied to cutting zone at a pressure of 5 kg/cm^2. The authors compared the performance of cryogenic machining with dry machining without using any coolant. Although the cryogenic cooling reduced the cutting forces by about 15% and flank wear by about 38%, it needed a large quantity of liquid nitrogen, which increase the overall cost of machining. The authors concluded that cryogenic cooling can be advantageous only in high speed and high feed machining.

Sreejith and Ngoi (2000) have listed one more method for achieving cooling in dry machining—method of thermoelectric refrigeration. It is based on Peltier effect. Peltier effect is the cooling of one junction and the heating of the other when electric current is maintained in a circuit consisting of two dissimilar conductors. This effect is even stronger in circuits containing dissimilar semiconductors. There can be many combinations of dissimilar metals or semiconductors, for example, copper and bismuth. However, method of thermoelectric refrigeration is not a popular method in machining.

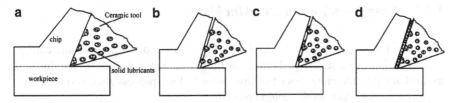

Fig. 3.5 Formation of self-tribofilm between the tool–chip interface in machining process. With permission from Jianxin et al. (2007). Copyright (2007) Elsevier

3.3.2 Lubrication in Dry Machining

One of the most prominent ways of achieving lubrication in dry machining is to use soft coatings on the cutting tools. These coatings are called self-lubricating coatings (Weinert et al. 2004). Typical self-lubricating coating is MoS_2, which is applied on a hard coating using DC sputtering technique. Kustas et al. (1997) deposited multilayer nanocoatings (100 bilayers of $13\text{Å}B_4C/18\text{Å}W$) on cemented WC-Co tools and HSS drills using balanced magnetron sputtering process. Dry machining on an AlSl 4140 steel (302 BHN) at 105 m/min showed a reduction in wear and consequent improvement in tool life compared to uncoated as well as $TiC–Al_2O_3–TiN$ trilayer coating. Dry drilling tests with a solid-lubricant, multi-layer-coated (MoS_2/Mo) (400 bilayers, bilayer thickness of 80Å and 3.2-µm total coating thickness) HSS drills on a Ti–6Al–4V alloy showed a reduction in torque (33%) and noseizure while seizure took place with the uncoated drill. Koshy (2008) deposited multilayer films of CrN and Mo_2N, where the Mo_2N forms a sacrificial oxide that lubricates at high temperature. The reduction of friction at high temperature points to a thermally activated self-lubricating mechanism in operation. As the coating wears the multilayer arrangement ensures the constant supply of lubricating oxides (from the Mo_2N layer) at the sliding interface.

Many researchers have incorporated solid lubricants in the ceramic matrix to develop the self-lubricating ceramic composites.

Self-lubricating ceramic composites consists of a supporting ceramic matrix surrounding the dispersed pockets of one or more softer lubricating species. Partially stabilized zirconia (PSZ) with a thin film (about 50 nm) of CuO can make a good self-lubricating tool.

Jianxin et al. (2007) produced Al_2O_3/TiC ceramic composites with the additions of CaF_2 solid lubricants by hot pressing. Tools made of these composites were used in the dry machining of hardened steel. Scanning electron microscopy revealed that a self-tribofilm was consistently formed on the wear surfaces and the composition of the self-tribofilm was found to be mainly CaF_2 solid lubricants.

In machining processes, the chip slides against the tool rake face at high speed, and induces high cutting temperature. Under such high cutting temperature, the CaF_2 solid lubricants may be released and smeared, and create a thin lubricating film on the rake face, thus resulting in a decrease in the friction coefficient. Figure 3.5 illustrates the schematic diagram of the formation process of self-tribofilm on the rake surface in machining processes.

3.3.3 Chip Handling in Dry Machining

The main problem with dry machining is the removal of chips from the cutting zone. This is a more severe problem in drilling. One way is to enlarge the flutes of the drill and provide more space to chips. In the milling process also, a cutter with a large spacing between teeth should be used.

Dry machining also requires some changes in machine tool for easy chip disposal. The chip handling and chip transportation are facilitated by a totally enclosed working area and an automatic chip conveyor. Some exhaust system can be fitted to remove the chips.

3.4 Some Examples of Implementation of Dry Machining

In their keynote paper on dry machining, Klocke and Eisenblatter (1997) provided a wide range of examples of successful implementation of dry machining of cast iron, steel, aluminum, super alloys, and titanium. Cast materials can be easily dry machined. Cast iron is very much suitable for dry machining because it generates discontinuous short chips, requires low forces, generates less temperature, and provides lubrication due to embedded graphite. Among various machining processes, drilling, reaming and tapping are less amenable to dry machining. In drilling of steel, there is a tendency to get jammed in the hole. One way of overcoming this tendency is to provide more drill taper toward shank. Dry machining shows positive effects in milling, where a tool carries out interrupted cutting. In that case, wet machining provides thermal shock and reduces the tool life. Many a times, it is necessary to use special tool geometries having provision of chip breakers and/or enhanced rake angle to reduce friction.

There are a number of specially designed workpiece material that provided enhanced machinability (Klocke and Eisenblatter 1997; Byrne et al. 2003). Ca-treated steels provide self-lubrication. Ca treatment generally leads to a conversion of highly abrasive alumina inclusion into more ductile Ca aluminates.

Dry machining of laminar grey cast iron at high cutting speeds using ceramic and CBN has been carried out. CBN is a more suitable cutting tool material for dry machining due to its low coefficient of thermal expansion and high thermal conductivity.

Dry machining is not efficient for conventional machining of superalloys and titanium alloys in most cases, but it is efficient for high speed milling. Non-ferrous material like aluminum alloy also is difficult to dry machine due to high coefficient of thermal expansion of aluminum alloys. For dry drilling, reaming, tapping or end-milling, it is essential to use tools with suitable cooling system or use MQL. For dry machining of aluminum, diamond-coated tools are best suited as they do not produce any built-up edge due to little affinity between carbon and aluminum.

So far the success of dry machining has been dependent on tool–workpiece combination. Byrne et al. (2003) has compiled the following examples of dry machining:

- Drilling of free cutting steel and cast iron
- Milling and turning of aluminum, steel, and cast iron
- Gear milling of steel and cast iron
- Broaching of steel and cast iron

3.5 Conclusion

In this chapter, a brief review of dry machining has been presented. Dry machining in this chapter has been defined as the machining without the use of cutting fluid. Sometimes, the dry machining can be employed without any additional arrangement of cooling such as in high speed machining, while sometimes additional arrangement for cooling the tool is employed. The additional arrangement does not pose any environmental hazard.

Some self-lubricating tools are being used in dry machining. For better chip disposal in dry machining, the cutting tool and machine tool require some modifications. This is essential for getting the optimum performance from dry machining.

References

Ber A, Goldblatt M (1989) The influence of temperature gradient on cutting tool's life. CIRP Ann Manuf Technol 38:69–73

Byrne G, Dornfield D, Denkena B (2003) Advanced cutting technology. CIRP Ann Manuf Technol 52(2):1–25

Hollis WS (1961) The application and effect of controlled atmospheres in the machining of metals. Int J Mach Tool Des Res 1:59–78

Jen TC, Gutierrez G, Eapen S, Barber G, Zhao H, Szuba PS, Labataille J, Manjunathaiah J (2002) Investigation of heat pipe cooling in drilling applications. part I: preliminary numerical analysis and verification. Int J Mach Tools Manuf 42:643–652

Jianxin D, Tongkun C, Lili L (2007) Self-lubricating behaviors of Al_2O_3/TiB_2 ceramic tools in dry high-speed machining of hardened steel. J Eur Ceram Soc 25:1073–1079

Kalyan Kumar KVBS, Choudhury SK (2008) Investigation of tool wear and cutting force in cryogenic machining using design of experiments. J Mater Process Technol 203:95–101

Klocke F, Eisenblatter G (1997) Dry cutting. CIRP Ann Manuf Technol 46(2):519–526

Koshy RA. (2008) Thermally activated self-lubricating nanostructured coating for cutting tool applications. Ph.D. Thesis, Northwestern University

Kustas FM, Fehrehnbacher LL, Komanduri R (1997) Nanocoatings on cutting tools for dry machining. CIRP Ann Manuf Technol 46:39–42

Landgraf G (2004) Dry goods. Cutting Tool Eng 56

Paul S, Chattopadhyay AB (1995) Effects of cryogenic cooling by liquid nitrogen jet on forces, temperature and surface residual stresses in grinding steel. Cryogenics 35:515–523

Paul S, Dhar NR, Chattopadhyay AB (2001) Beneficial effects of cryogenic cooling over dry and
    wet machining on tool wear and surface in turning AISI 1060 steel. J Mater Process Technol
    116:44–48
Sanchez LEA, Scalon VL, Abreu GGC (2011) Cleaner machining through a toolholder with
    internal cooling. Proceedings of third international workshop on advances in cleaner produc-
    tion, São Paulo, Brazil, May 18–20, 2011, pp 1–10
Sreejith PS, Ngoi BKI (2000) Dry machining: machining of the future. J Mater Process Technol
    101:287–291
Yildiz Y, Nalbant M (2008) A review of cryogenic cooling in machining processes. Int J Mach
    Tools Manuf 48:947–964
Wang ZY, Rajukar KP (2000) Cryogenic machining of hard to cut materials. Wear 239:168–175
Weinert K, Inasaki I, Sutherland JW, Wakabayashi T (2004) Dry machining and minimum
    quantity lubrication. CIRP Ann Manuf Technol 53(2004):511–537
Zhao H, Barber GC, Zou Q (2002) A study of flank wear in orthogonal cutting with internal
    cooling. Wear 253:957–962

Chapter 4
Gas-Cooled Machining

Abstract It is possible to use coolant and/or lubricant in gas form instead of solid or liquid form. Cooling action in gas-cooled machining is mainly through convection. The heat transfer by the gas jet can be controlled by controlling the jet velocity. Gases also provide inert atmosphere and lubrication. There is no coolant disposal problem in gas-cooled machining as the gases get merged into the atmosphere. These gases are already present in the atmosphere and are harmless. In this chapter, typical gas-cooling systems are described.

Keywords Air-cooled machining • Cutting force • Cutting temperature • Dry turning • Gas-cooled machining • Surface roughness • Tool life • Tool wear • Vortex tube

4.1 Introduction

The recent challenge before the researchers is to do machining in a green environment without sacrificing the machining performance. The application of gas cooling in machining is one way of achieving the conflicting goals of machining performance and green environment. Wet machining and minimum quantity of lubrication (MQL) machining pose problem of disposal of the cutting fluid. MQL often produces the mist. The mist in the industrial environment can have serious respiratory effects on the operator. Gas-cooled machining has no such adverse effect on human beings and as such it may work as an alternative to liquid coolant in machining. Air-cooled systems are a special case of gas-cooled systems. Air is a natural resource and it is readily available everywhere. The only processes required in air-cooling system are to compress, dry, and discharge the air with a set pressure. In most of the factories, the compressed air is already available for doing various types of tasks. Thus, no additional arrangements are required. In this chapter, some gas cooling applications are described.

U.S. Dixit et al., *Environmentally Friendly Machining*, SpringerBriefs
in Applied Sciences and Technology, DOI 10.1007/978-1-4614-2308-9_4,
© Springer Science+Business Media, LLC 2012

4.2   Typical Gases Used in Machining

The earliest application of gas coolant goes back to 1930s, when nitrogen gas was applied as a coolant (Shaw 2002). Nitrogen gas formed a film on tool surface, chip easily moved away from tool face, and lower friction occurred in tool–workpiece–chip interface. With nitrogen gas the tool life increased considerably. The use of nitrogen gas in high speed milling of Ti-6Al-4V enhanced the machining performance (Ying-lin et al. 2009). Other gases that have been used are carbon dioxide, argon, water vapor, and air. Some researchers have observed that cooling by oxygen can enhance tool life, whereas in some cases it reduced tool life due to oxidation effect (Rowe and Smart 1963; Johansson and Lindström 1971). It has been observed that use of oxygen reduces the chip–tool contact length. In many cases, oxygen prevents the formation of built-up edge by oxidizing the chip surface. Çakir et al. (2004) observed that in the turning of AISI1040, oxygen has more lubricating effect than nitrogen and CO_2 has more lubricating effect than oxygen.

Hollis (1961) carried out turning using CO_2 and CO_2 along with argon gas. Argon provides an inert atmosphere and provides more enhanced tool life and less hardening at the machined surface. Particularly, in the machining of titanium, argon prevents the absorption of CO_2 on the machined surface. In the 1990s, Podgorkov and Godlevski proposed a new and pollution-free cutting technique with water vapor as coolant and lubricant during cutting process. They used water vapor with temperature less than 100°C. The coolant was a mixture of water in gas and liquid form. Water vapor is cheap and pollution free, thus an ideal coolant. It has been observed that water vapor lubrication causes more uniform cooling and increases the tool life of cemented carbide tools by about 1.5 times in turning of stainless steel.

Williams and Tabor (1977) have suggested a model for the lubricating action of gas or vapor during orthogonal machining based on their experimental study. According to their model, a number of capillaries get formed at the interface of chip and rake face of the tool. The lubricant in gas or liquid form is drawn in these capillaries and creates a boundary lubrication layer. A typical capillary is shown in Fig. 4.1. The cross-sectional dimensions of the capillary are a and ma and the length is l. The cross-sectional dimensions are of submicron size. Lubricating action of the cutting fluid requires that the cutting fluid absolutely penetrates into the capillary and penetration time has to be less than the capillary lifetime. The difference of capillary lifetime and penetration time is called the storing time (Junyan et al. 2010). The longer the penetration time, better the performance of the lubricant. The penetration time of gas is less by an order of magnitude compared to that of liquid. Hence, while using gas, the storing time is increased and the lubricating action is enhanced.

Fig. 4.1 A schematic representation of single interfacial capillary. With permission from Williams and Tabor (1977). Copyright (1977) Elsevier

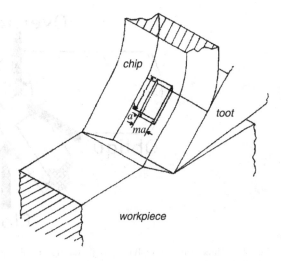

4.3   Methods of Applying Gas Jets

There are various ways of applying the gas jet. For example, Ko et al. (1999) have considered three ways of applying the gas coolant as shown in Fig. 4.2. It is also possible to use multiple jets. When the jet comes from the top, it cools the top surface of the chip and helps in its curling. The tool–chip contact length is reduced and the friction is minimized. When the gas jet impinges on the flank surface, it carries away the heat generated due to rubbing of the flank surface with the work.

Hilsch (1947) has described the use of a vortex tube for cooling of the air. It was first described by Ranque. In this arrangement (Fig. 4.3), compressed air enters tangentially to a cylindrical chamber. A turbulent flow of gas in a screw-like motion escapes through both ends. This rotating air stream produces a region of increased pressure near the wall inside the cylinder, and a region of decreased pressure near the axis. One end of the cylinder is closed by a diaphragm which permits the escape of air only from the central region. The other end allows the flow of axial portion of the air. The air entering through the central diaphragm shows a reduced temperature, while the air escaping through the other end shows a temperature increase. If the temperature of the compressed air coming out from the nozzle is T_a, the cold air temperature is T_c, and the hot air temperature is T_h, then the following approximate relation is valid:

$$\text{(Mass flowrate of cold air)} \times (T_a - T_c)$$
$$= \text{(Mass flowrate of hot air)} \times (T_h - T_a). \tag{4.1}$$

This is because the amount of heat given to the heated air must be equal to the amount of heat removed from the cooled air.

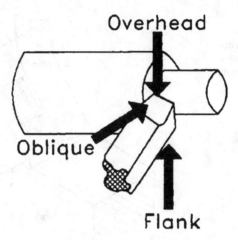

Fig. 4.2 Three ways of applying gas jet during machining. From Ko et al. (1999). Copyright (1999) Springer

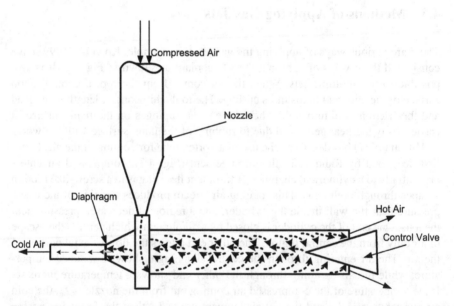

Fig. 4.3 Schematic of vortex tube

Ko et al. (1999) has used the vortex tube for supplying air jet to a turning process. The compressed air at 5 bar enters the cylindrical chamber. The part of gas, mostly the axial one, escapes through the larger tube whose exit is at a distance of about 30 times the diameter of the tube. By this distance, the air loses its screw-like motion. The fraction of the air through this end is controlled by a valve. The remaining air escapes from the other end through the center portion of a diaphragm. The temperature of this air is about 20°C less compared to the ambient temperature.

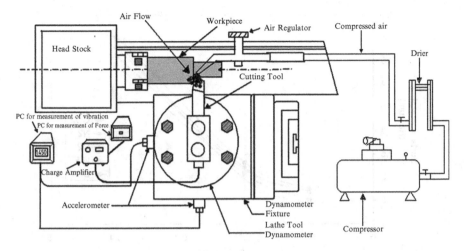

Fig. 4.4 Schematic drawing of dry compressed air used in turning operation (Sarma 2009)

Figure 4.4 shows a schematic of the dry compressed air used in turning operation (Sarma 2009). Compressed air was generated in a single stage compressor and delivered through a drier to the nozzle. The velocity of airflow was approximately 150 m/s and air temperature was about 2°C less than the ambient temperature. The jet is impinged on the tool–work interface at about 45° to the longitudinal direction for discharging the air. This position of the jet was found to be optimum experimentally.

4.4 Performance of Gas Coolants

A limited number of publications have been found toward application of gas cooling in machining. Sharma et al. (2009) have presented a good overview of major advances in cooling techniques in machining for improvement of productivity where compressed air cooling is one of the major applications. Beside compressed air, sometimes other gases such as nitrogen gas or liquid nitrogen were delivered under pressure and directed into the tool–chip interface.

Çakir et al. (2004) applied several gases (nitrogen, oxygen, and carbon dioxide) in turning of AISI1040 steel with carbide tool, which produced lower cutting force compared to dry and wet cutting. The value of surface roughness was found nearly equal in all the gases (Fig. 4.5). Among all the gases, CO_2 provides slightly better surface roughness. Gas application in turning provided higher shear angle value compared to dry and wet turning at lower feeds. The higher shear angle implies the larger cutting ratio (ratio of uncut chip thickness to chip thickness). Thus, there is less chip thickening and cutting is more efficient. This invariably means the less cutting forces.

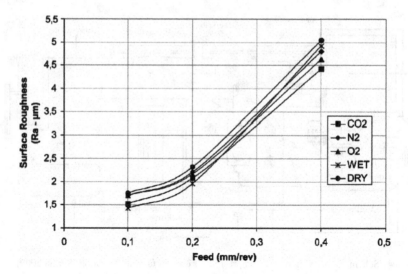

Fig. 4.5 Effect of feed on surface roughness under various lubrication conditions. With permission from Çakir et al. (2004). Copyright (2004) Elsevier

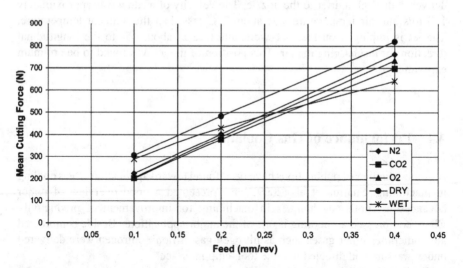

Fig. 4.6 Variation of mean cutting force with feed under various lubrication conditions. With permission from Çakir et al. (2004). Copyright (2004) Elsevier

Figure 4.6 shows the variation of mean cutting force with feed. At lower feed, the mean cutting forces are lower for gas-cooled machining than for dry or wet machining. At a feed of 0.4 mm/rev, wet machining provides the lowest force. Among gases, the performance of CO_2 is the best in reducing the mean cutting force.

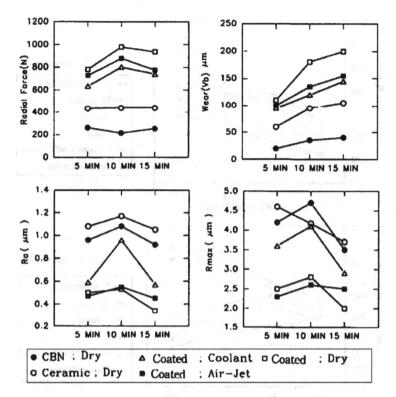

Fig. 4.7 Performance of various cutting tools under different lubrication conditions. From Ko et al. (1999). Copyright (1999) Springer

Altan et al. (2002) used oxygen gas instead of cutting fluid in machining. The experimental study concluded that the application of oxygen improved surface quality and reduced cutting forces. It was observed that gas pressure also influenced as a chip breaker. Ko et al. (1999) developed an air-cooling system with air-vortex flow arrangement for reducing the heat generated at the tool–chip interface during the turning of heat-treated (SAE 52100) bearing steel. Initially, air was injected in the cutting region at 0°C and tool wear was compared with dry cutting. The tool wear was found less in air cooling compared to the dry cutting. However, the performance in terms of tool wear was not superior to the wet cutting.

Figure 4.7 shows that air-cooled TiN coated tools provide better surface finish in hard turning, although they provide higher cutting force and wear compared to CBN and ceramic tools. However, CBN and ceramic tools are expensive and considering this fact use of TiN coated carbide tool with air-cooling may be more economical.

Stanford et al. (2009) investigated the use of gaseous and liquid nitrogen as a cutting fluid while turning of plain carbon steel with uncoated turning tools. They conducted several experiments considering different cutting environments out of which compressed air blast was one of the important environments. Figure 4.8a shows that the tool crater wear is the lowest with fluid cooling. Figure 4.8b shows

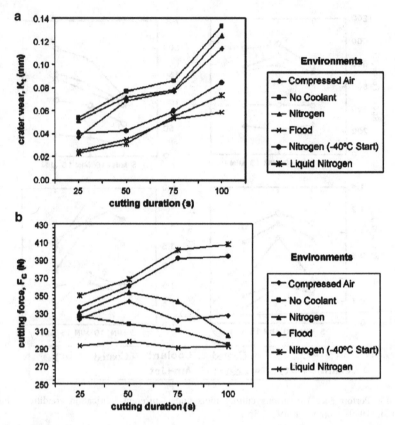

Fig. 4.8 Details of (**a**) crater wear, and (**b**) cutting force, for cutting En32b, uncoated tooling, cutting speed = 400 m/min, feed = 0.12 mm/rev, depth = 1.2 mm. From Stanford et al. (2009). Copyright (2009) Elsevier

that the cutting force is the lowest with liquid nitrogen cooling. Compressed air cooling provides an intermediate performance with respect to both cater wear and cutting forces. Figure 4.9 shows that the compressed air cooling reduces the temperature during machining. Considering that the cost of compressed air is not very high, it can become a reasonable substitute of flood coolant.

Sarma and Dixit (2007) studied the performance of the dry and air-cooled turning of grey cast iron with a mixed oxide ceramic cutting tool. The air cooling was achieved by a jet of compressed air. In their study, it was reported that at high cutting speeds, the air cooling reduces the flank wear but this is not the case at low cutting speeds. In fact, at low cutting speed, for some combination of cutting parameters, the air cooling slightly increased the flank wear. At low cutting speed, the heat generation is low and tool-job temperature is not as high as in high-speed turning. Therefore, the forced convective heat transfer is not as effective as in high-speed turning. Instead, sometimes in dry turning at low speed, a slight increase of job temperature causes a slight reduction in tool wear

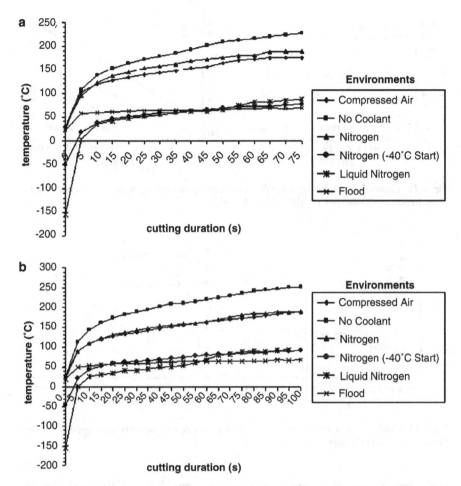

Fig. 4.9 Remote cutting temperature profiles for cutting En32b, uncoated tooling, cutting speed = 400 m/min, feed = 0.12 mm/rev, depth of cut = 1.2 mm (**a**) cutting duration 75 s and (**b**) cutting duration 100 s. From Stanford et al. (2009). Copyright (2009) Elsevier

due to softening of the material. Compressed air plays a vital role in reducing the temperature at high-speed turning of grey cast iron with ceramic tool. Further, there is very slight reduction in surface roughness in air-cooled cutting in comparison to dry turning up to a speed of 400 m/min. However, at a cutting speed of 480 m/min, where the dry turning provides a very poor surface finish due to rapid tool wear, the air cooling provided a very good surface finish. In all the cases, air-cooled turning lowered the cutting and feed forces as compared to corresponding forces in dry turning. Thus, air-cooled turning seems to be a good environmentally friendly option for high speed turning.

Junyan et al. (2010) have studied the lubrication action of water vapor in the turning of ANSI 304 stainless steel by carbide tool. It is observed that the friction

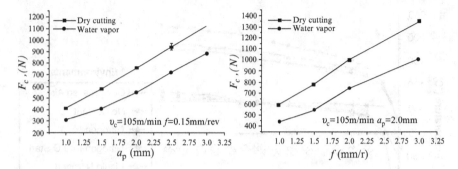

Fig. 4.10 Comparison of cutting force F_c in dry cutting and cutting with water vapor cooling. From Junyan et al. (2010). Copyright (2010) Elsevier

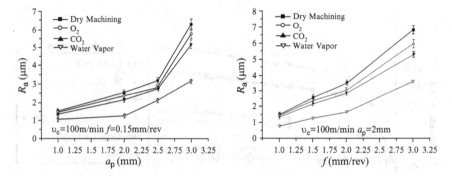

Fig. 4.11 Comparison of surface roughness obtained under various lubrication condition. From Junyan et al. (2010). Copyright (2010) Elsevier

coefficient is lowered in cooling with water vapor. Figure 4.10 shows that the cutting force is lower with water vapor cooling compared to dry cutting. Figure 4.11 shows that the surface roughness is the lowest with water vapor cooling.

4.5   Conclusion

In this chapter, the use of gases as coolant is described. It is noted that the performance of gas coolants is better than the liquid coolant in many cases. Moreover, there is no waste disposal problem as the gases can be released to atmosphere. Among the gases, compressed air seems to be most viable alternative. It is readily available in most of the factories and does not cause environmental pollution at all. In Chap. 5, a detailed comparison of dry turning and compressed air-cooled turning has been presented.

References

Altan E, Kiyak M, Cakir O (2002) The effect of oxygen gas application into cutting zone on machining. Proceedings of the sixth biennial conference on engineering systems design and analysis, Istanbul, pp 1–5

Çakir O, Kiyak M, Altan E (2004) Comparison of gases applications to wet and dry cuttings in turning. J Mater Process Technol 153–154:35–41

Hilsch R (1947) The use of the expansion of gases in a centrifugal field as cooling process. Rev Sci Instrum 18:108–113

Hollis WS (1961) The application and effect of controlled atmosphere in the machining of metals. Int J Mach Tools Des Res 1:59–78

Johansson U, Lindström B (1971) Tool wear by chatter caused by oxygen pumping. Ann CIRP 19:317–321

Junyan L, Huanpeng L, Rongdi H, Yang W (2010) The study on lubrication action with water vapor as coolant and lubricant in cutting ANSI 304 stainless steel. Int J Mach Tools Manuf 50:260–269

Ko JT, Kim HS, Chung BG (1999) Air-oil cooling method for turning of hardened material. Int J Adv Manuf Technol 15:470–477

Podgorkov VV (1992) Method of cutting in application. Patent of USSR 1549721 MCI B23Q (in Russia)

Podgorkov VV, Kapustin AS, Godlevski VA (1998) Water steam lubrication during machining. Tribologia 29:890–901 (in Polish)

Rowe GW, Smart EF (1963) The importance of oxygen in dry machining of metal on lathe. Br J Appl Phys 14:924–926

Sarma DK (2009) Experimental study, Neural network modeling and optimization of environment-friendly air-cooled and dry turning processes. Ph.D. Thesis, IIT Guwahati

Sarma DK, Dixit US (2007) A comparison of dry and air-cooled turning of grey cast iron with mixed oxide ceramic tool. J Mater Process Technol 190:160–172

Sharma VS, Dogra M, Suri NM (2009) Cooling techniques for improved productivity. Int J Mach Tools Manuf 49:435–453

Shaw MC (2002) Metal cutting principles. CBS Publishers & Distributors, Delhi

Stanford M, Lister PM, Morgan C, Kibble KA (2009) Investigation into the use of gaseous and liquid nitrogen as a cutting fluid when turning BS 970-80A15 (En32b) plain carbon steel using WC-Co uncoated tooling. J Mater Process Technol 209:961–972

Williams JA, Tabor D (1977) The role of lubricants in machining. Wear 43:275–292

Ying-lin KE, Hui-yue D, Gang L, Ming Z (2009) Use of nitrogen gas in high-speed milling of Ti-6Al-4V. T Nonferr Metal Soc China 19:530–534

Chapter 5
A Detailed Comparison of Dry and Air-Cooled Turning

Abstract Turning is one of the most widely used machining processes. In this chapter, a detailed comparison of dry and air-cooled turning has been carried out based on the experimental study. Machining of cast iron and steel is considered. Cutting tools considered in the study are TiN-coated carbide, CBN, and ceramic. It is observed that air cooling always provides better machining performance. It is particularly needed in the high speed machining and hard turning.

Keywords Air-cooled turning • Carbide cutting tools • CBN cutting tool • Ceramic cutting tool • Cutting force • Dry turning • Grey cast iron • Surface roughness • Tool life • Tool wear

5.1 Introduction

Turning is one of the most widely used machining processes. The study of turning process may provide an insight to other machining processes as well. Sarma (2009) has carried out an extensive investigation on the air-cooled turning processes. A jet of compressed air was used in place of coolant in the machining with TiN-coated carbide, ceramic, and CBN cutting tools. The air was only slightly cooler than ambient temperature. No special attempt was made to cool the air further. The heat removal was due to increased convective heat transfer coefficient at higher jet velocity. The results indicate a good performance of air-cooled turning. In the subsequent sections, some results from Sarma (2009) are discussed.

5.2 Turning of Mild Steel with Coated Carbide Tool

Sarma (2009) carried out dry and air-cooled turning of mild steel with TiN-coated carbide tools. Rolled mild steel containing 0.35% carbon was used for turning. The hardness of workpiece was 130 BHN, yield strength 290 MPa, and ultimate

U.S. Dixit et al., *Environmentally Friendly Machining*, SpringerBriefs
in Applied Sciences and Technology, DOI 10.1007/978-1-4614-2308-9_5,
© Springer Science+Business Media, LLC 2012

tensile strength 477 MPa. Coated carbide insert was squared shaped and ASA tool signature was (-1)-(-1)-6-6-45-45-1.6. In air-cooled turning, the air was dried and delivered through a nozzle of diameter 5 mm at a pressure of 2 bars. The velocity of air flow was approximately 150 m/s and air temperature was about 2°C less than the ambient temperature. The air jet impinged on the tool–work interface at about 45° to the longitudinal direction.

Figure 5.1 compares the surface roughness for dry and air-cooled turning for eight different cutting conditions. Over 100 mm of machined length, the surface roughness was measured at 9–12 places and the mean values were calculated. These are plotted in Fig. 5.1. Figure 5.1a–d represents the machining at lower speed and Fig. 5.1e–h represents the machining at higher speed. In most of the cases, surface finish is found to be almost equal for both dry and air-cooled turning in all the cutting conditions. In some cases, there is a difference in surface roughness obtained in air-cooled and dry turning values. Sometimes, dry turning has provided better surface finish. Increased temperature in metal cutting softens the material, but reduces the hardness of the tool. These have two opposing effects on metal cutting performance in general and surface roughness in particular. The net result depends on the relative influence of these opposing factors.

Sarma (2009) has carried out the air cooling at fixed temperature and velocity of the air. The effect of lowering the air temperature and changing the jet velocity was not studied. It may be possible to obtain the better machining performance by optimizing the parameters of the air. For example, Choi et al. (2002) carried out the grinding of a spindle shaft material having 58–60 HRC hardness with white alumina and CBN grinding wheel. They compared the grinding performance of flood coolant and compressed cold air. Four levels of air temperature $(-4°C, -10°C, -15°C,$ and $-25°C)$ and two levels of air velocity (40 and 80 m/s) were considered. In general, a better surface finish of the ground surface was obtained at lower air temperature and higher jet velocity. The cold air was produced by a vortex tube.

Figure 5.2 compares the progression of flank wear for dry and air-cooled turning. It is observed that at high cutting speeds (Fig. 5.2e–h), air has played a significant role in reducing the flank wear. However, at low cutting speeds (Fig. 5.2a–d), many a times, air cooling causes more flank wear. At low cutting speed, heat generation is low and the effect of air cooling is marginal. In fact, some increase in the temperature causes material softening and reduces the tool wear. In turning of mild steel with coated carbide tool, no crater wear was observed during dry turning at high cutting speed.

Figures 5.3 and 5.4 depict the variation of cutting force and feed force, respectively, with respect to cutting time for eight different cutting conditions. It is observed that in most cases, the value of cutting force or feed force is almost equal for both air-cooled and dry turning. Thus, no special effect has been seen for air-cooled turning. Comparing the cutting force with feed force for same cutting condition, the cutting force is found more than the feed force. It is a general phenomenon that cutting force is higher than the feed force.

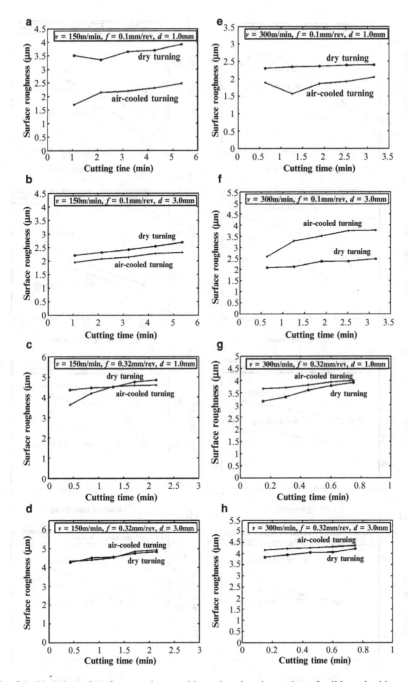

Fig. 5.1 Variation of surface roughness with cutting time in turning of mild steel with coated carbide tool (Sarma 2009)

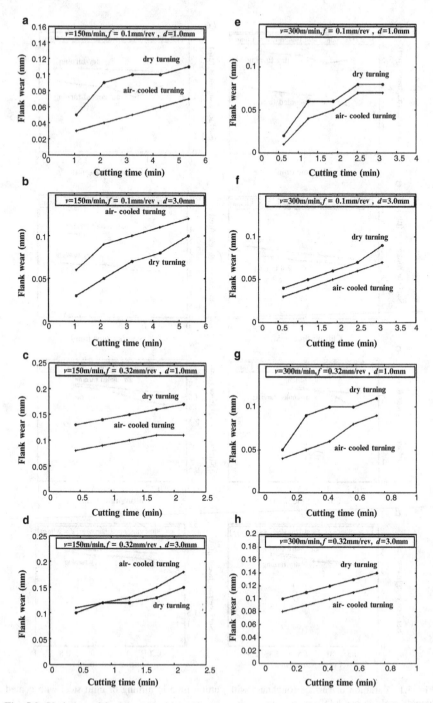

Fig. 5.2 Variation of flank wears with cutting time in turning of mild steel with coated carbide tool (Sarma 2009)

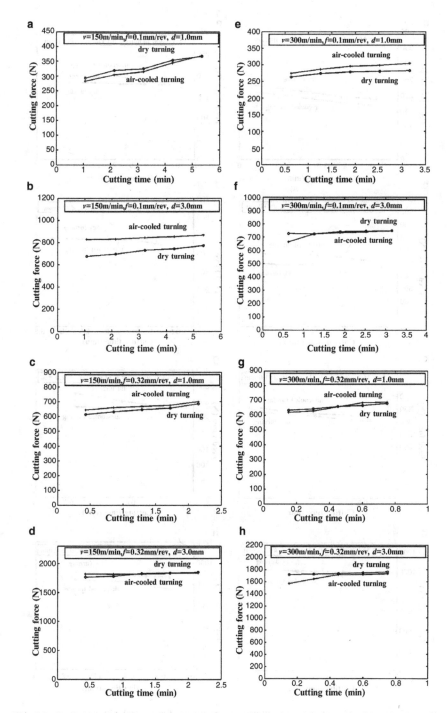

Fig. 5.3 Variation of cutting force with cutting time in turning of mild steel with coated carbide tool (Sarma 2009)

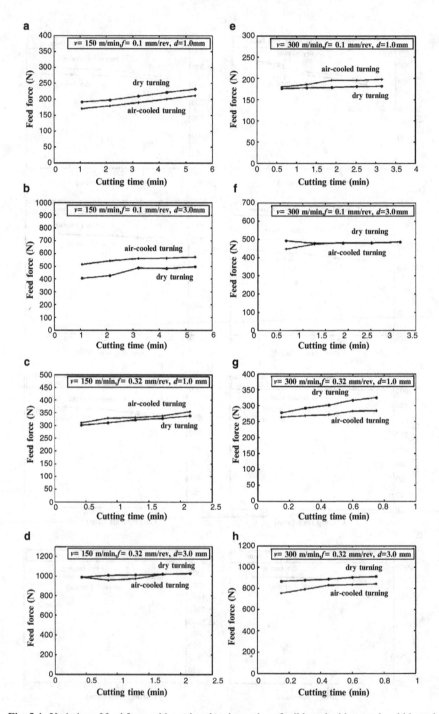

Fig. 5.4 Variation of feed force with cutting time in turning of mild steel with coated carbide tool (Sarma 2009)

5.3 Turning of Grey Cast Iron with Ceramic Tool

Alumina-based cutting tools are attractive alternatives to carbide tools for the machining of steel and cast iron because of their hot hardness (Sadasivan and Sarathy 1999). Two main types of these ceramics are pure/white oxide ceramic and mixed/black oxide ceramic. White oxide ceramic containing Al_2O_3 with sintered additives and without metallic binder phase is relatively brittle. Its toughness can be improved by embedding fine zirconia (ZrO_2) particles by an amount of 3–5% into the aluminum oxide matrix. Such a ceramic is called dispersion ceramic. White oxide ceramic is used in rough machining of grey cast iron, nodular cast iron, and chilled cast iron. Besides aluminum oxide, the black oxide ceramic contains titanium oxide and/or titanium carbonitride in the order of about 30% by weight. It is generally used for machining of hard materials and finish machining of cast iron.

Sarma and Dixit (2007) studied the performance of finish turning of a grey cast iron with mixed oxide ceramic tool. Chemical composition by weight of the grey cast iron was C3.2%–Si1.8%–Mn0.36%–S1.8%–P0.05%. Its hardness was 143 BHN, tensile strength 86 MPa, and compressive strength 512 MPa. First, a series of experiments were conducted for ceramic tool in the speed range of 480–600 m/min as per Table 5.1. The machining was carried out in both dry and air-cooled conditions. Each test was conducted with a new cutting edge and machining was stopped at a cutting length of 100 mm. The time was recorded for each cut. The vibration signals and cutting forces were measured during cutting operation. The maximum flank wear and surface roughness were measured after completion of each pass.

Figure 5.5 shows the flank wear vs. cutting time for dry turning and air-cooled turning in the first five operating conditions of Table 5.1. During dry turning, crater wear was observed in the last three operating conditions i.e., in experiment numbers 6, 7, and 8, which led to early tool breakage. The surface finish was also found to be very poor at these conditions and a large amount of heat was generated. However, in air-cooled turning, no crater wear was seen in experiment number 7 and combination of crater wear and flank wear were seen in experiment numbers 6 and 8.

Table 5.1 Operating conditions in turning of grey cast iron with ceramic tool at high cutting speed (Sarma 2009)

Experiment No.	Cutting speed (m/min)	Feed (mm/rev)	Depth of cut (mm)
1	480	0.04	0.5
2	600	0.04	1.5
3	600	0.32	1.5
4	480	0.04	1.5
5	480	0.32	1.5
6	600	0.32	0.5
7	600	0.04	0.5
8	480	0.32	0.5

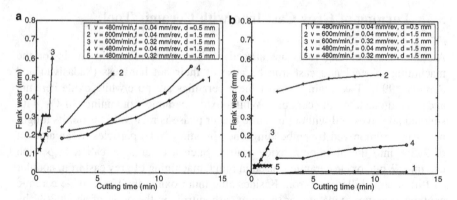

Fig. 5.5 Progression of tool flank wear with cutting time (**a**) dry turning (**b**) air-cooled turning of grey cast iron with ceramic tool at high cutting speed. With permission from Sarma and Dixit (2007). Copyright (2007) Elsevier

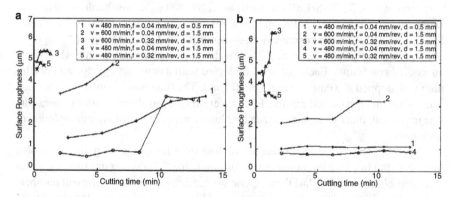

Fig. 5.6 Progression of surface roughness with cutting time (**a**) dry turning (**b**) air-cooled turning of grey cast iron with ceramic tool at high cutting speed. With permission from Sarma and Dixit (2007). Copyright (2007) Elsevier

It indicates that air cooling plays a vital role in reducing the temperature at high-speed turning on ceramic tool. Both surface finish and tool life were found to be satisfactory in experiment numbers 6, 7, and 8 during air-cooled turning (not shown in figure). For the first five operating conditions, air-cooled turning would provide better tool life as is evident from Fig. 5.5a,b. Also, during dry turning, for conditions 2, 3, and 5, tool breakage occurred after 3–4 passes. However, in air-cooled turning, the tool breakage was not observed up to even 6 passes (each pass consisting of a length of 100 mm), after which the experiment was stopped.

Figure 5.6 shows the variation of surface roughness of the turned component with cutting time for dry and air-cooled turning. It is observed that the surface roughness at operating conditions 3 and 5 is quite high in both dry and air-cooled turning.

These cases pertain to a high feed of 0.32 mm/rev. For the cases, 1, 2, and 4, air-cooling reduces the surface roughness significantly.

For the conditions 1 and 4, which pertain to cutting speed of 480 m/min and feed of 0.04 mm/rev, a very low and almost constant surface roughness was observed in air-cooled turning. Tool wear was also very low in these conditions. This shows that in high-speed turning, the air-cooling improves the surface roughness as well as tool life apart from the reduction in cutting forces.

Thus, it is seen that high speed turning of grey cast iron with mixed alumina ceramic tools provides a low tool life. This observation is in line with the observations of Ghani et al. (2002) for high speed turning of nodular cast iron by mixed alumina ceramic. However, air-cooled turning provides an enhanced tool life and may be adopted by industries for high speed machining by ceramic tools.

Figure 5.7 compares the surface roughness for dry and air-cooled turning for eight different cutting conditions in the medium speed range. Over a length of 100 mm, the surface roughness was measured at 9–12 places and the mean value was taken. Finally, the mean value of all replicates at a cutting condition was taken and the same have been plotted in Fig. 5.7. It is observed that, in general, the surface roughness is lower at higher cutting speeds compared to lower cutting speeds. However, at high speed, a combination of low feed and high depth of cut provided a relatively poorer surface finish (Fig. 5.7f) with air cooling. It is observed that air cooling does not improve the surface finish. For one condition (Fig. 5.7f), air cooling generated significantly greater surface roughness compared to dry turning. At this condition, which is low feed and high depth of cut, air-cooled turning produces more vibrations compared to dry turning.

Figure 5.8 compares the progression of flank wear for dry and air-cooled turning. It is noted that at high cutting speeds (Fig. 5.8e–h), the air cooling reduces the flank wear. This is not the case at low cutting speeds (Fig. 5.8a–d). In fact, at low cutting speed, for some cases, the air cooling slightly increases the flank wear. At low cutting speed, the heat generation is low and tool-job temperature is not as high as in high speed machining. Therefore, the forced convective heat transfer is not as effective as in high speed machining and one should not expect any significant reduction in tool wear due to reduction of tool-tip temperature. Instead, sometimes in dry turning at low speed, a slight increase of job temperature causes a slight reduction in tool wear due to softening of the material.

Figure 5.9 shows the variation of main (vertical) cutting force with the cutting time for eight cutting conditions. Interestingly, in some cases of low speed turning, the cutting force in dry turning was found greater compared to air-cooled turning, although the tool wear in dry-turning was less than that in air-cooled turning. In all the cases of high speed turning, the vertical cutting force is less in air-cooled turning due to less flank wear.

Figure 5.10 shows the variation of feed force with cutting time for eight cutting conditions. Here, also it is observed that forces in air-cooled turning are lower, although the difference is less pronounced than in the case of cutting force (Fig. 5.9). Unlike cutting force, feed force increases significantly with the progression of cutting particularly at high cutting speed, thus making it more suitable for

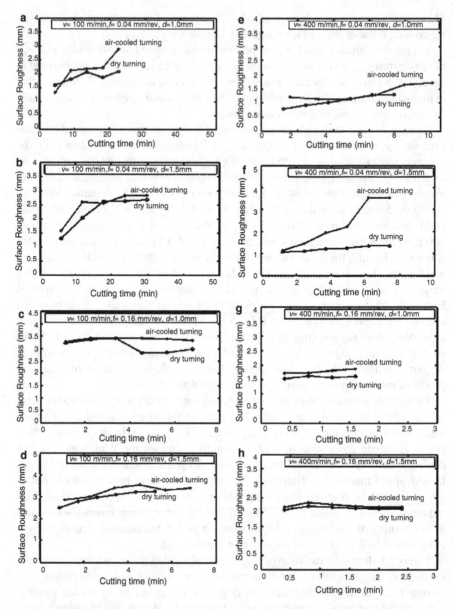

Fig. 5.7 Variation of surface roughness with cutting time in turning of grey cast iron with ceramic tool. With permission from Sarma and Dixit (2007). Copyright (2007) Elsevier

indirect measurement of tool wear. It is also noted that percentage increase of force due to tool wear is more at low feed than at high feed. This observation is in line with the theoretical analysis of Smithy et al. (2000), who showed that for a given tool and work–piece combination, the incremental increase in the cutting forces due

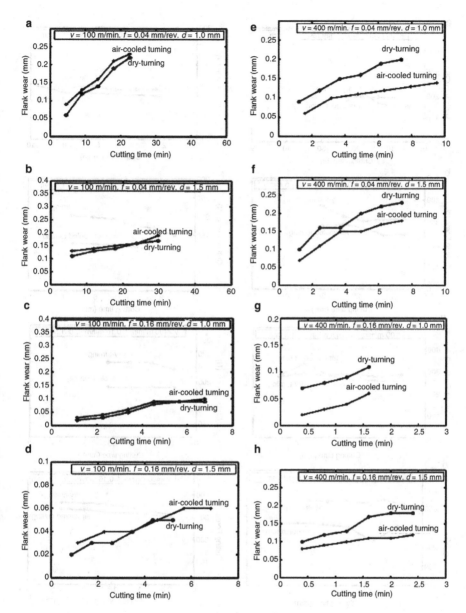

Fig. 5.8 Progression of tool flank wear with cutting time in turning of grey cast iron with ceramic tool. With permission from Sarma and Dixit (2007). Copyright (2007) Elsevier

to tool flank wear is solely a function of the amount and nature of the wear and is independent of cutting condition in which tool wear was produced. Thus, it may be a good strategy to carry out the indirect measurement of tool wear at low feeds, because the relative effect of wear is more prominent at low feeds.

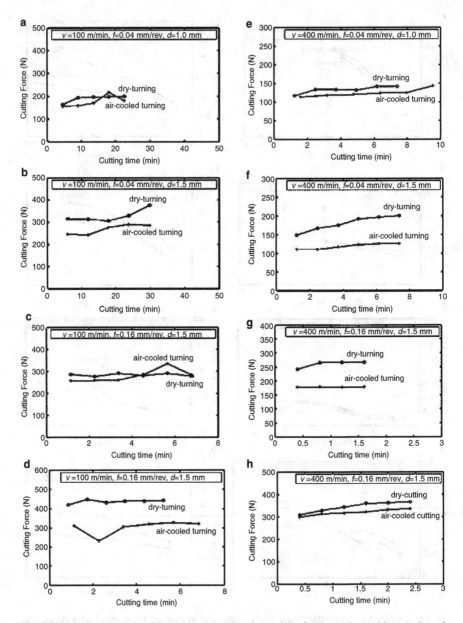

Fig. 5.9 Variation of cutting force with cutting time in turning of grey cast iron with ceramic tool. With permission from Sarma and Dixit (2007). Copyright (2007) Elsevier

The major advantage of air-cooled turning seems to be reduction of tool wear at high speed. Figure 5.11 shows the photographs of the progression of flank wear in dry and air-cooled turning for a cutting condition of speed 400 m/min, feed 0.04 mm/rev, and depth of cut 1 mm. It is seen that with air cooling, the growth of

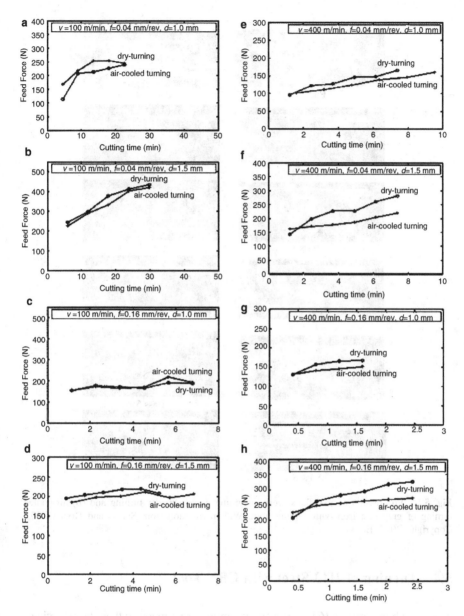

Fig. 5.10 Variation of feed force with cutting time in turning of grey cast iron with ceramic tool. With permission from Sarma and Dixit (2007). Copyright (2007) Elsevier

wear land is minimized. Other advantage is the lowering of cutting and feed forces. However, the impact of air-cooling on surface roughness is insignificant except at very high cutting speed (480–600 m/min range) as discussed earlier. Overall, air-cooled turning using ceramic tool offers an attractive alternative to dry turning.

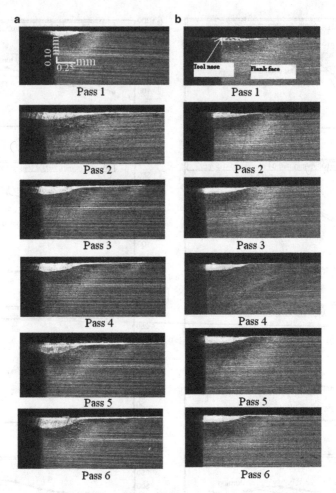

Fig. 5.11 Progression of tool flank wear in turning passes (**a**) dry turning and (**b**) air-cooled turning of grey cast iron with ceramic tool. With permission from Sarma and Dixit (2007). Copyright (2007) Elsevier

5.4 Turning of H13 Steel with CBN Tool

The cubic boron nitride (CBN) is a significantly harder material than the ceramic or coated carbide tool. It is used to carry out machining on hard materials due to its properties of chemical inertness, stability at high temperatures, hot hardness, etc. The basic material of CBN is hexagonal boron nitride (HBN), which is converted to CBN grits by using a solvent catalyst under high pressure and temperature. The structure of CBN does not change to HBN below 1,200°C at atmospheric pressure. CBN is used as a solid insert consisting of an upper layer of CBN laid onto a hard metal base (usually titanium nitride, TiN). Sometimes CBN is brazed on to a corner of a hard metal indexable insert. The main application of CBN tool is for turning of hardened steels of 45–68 HRC. Normally, these types of materials are ground rather than turned by a single point cutting tool. However, when large amount of material is to be removed, turning with CBN tool is economical. Turning of materials of hardness less than 45 HRC with CBN tool is not economical as carbide and ceramics can provide the same production rate without any significant difference in wear.

A number of experiments were carried out by Sarma (2009) on dry turning and air-cooled turning of AISI H13 hot work die steel with CBN tool. The diameter of the workpiece was 75 mm and the length was 300 mm. The job was supported at both ends. The workpiece was heat treated to 46 HRC. The cutting tool was made of a low CBN content material with 60% CBN along with a titanium nitride (TiN) phase. The manufacturer's catalog recommends the tool for finish hard turning of steels due to its high wear resistance and high chemical stability. The average hardness of the tool is $\geq 2,900$ HV_3 (Vickers hardness with 3-kg load). The ASA tool signature of this cutting tool was $(-6°)-(-6°)-6°-5°-12°-15°-0.4$.

The tool manufacturer recommended the following cutting conditions for machining of hardened steel of 45 HRC by CBN 7020:

- Cutting speed less than 250 m/min
- Feed in the range of 0.05–0.25 mm/rev
- Depth of cut in the range of 0.06–0.40 mm

Based on manufacturer's recommendations and review of the literature, the following ranges of the process parameters were chosen-cutting speed (v): 125–215 m/min, feed (f): 0.05–0.16 mm/rev, and depth of cut (d): 0.06–0.16 mm. It was decided to carry out the experiments with two-level full factorial design corresponding to three process parameters. The total number of full factorial experiments are eight, i.e., $2^3 = 8$.

Each test was conducted with a new cutting edge and machining was stopped at a cutting length of 100 mm. The time was recorded for each cut. For different experiments, the cutting time will be different depending upon the cutting speed and feed. The cutting forces in cutting direction and feed direction were measured during cutting operation. Maximum flank wear and surface roughness were measured after completion of every passes of machining. The tool rejection criteria were:

- Maximum flank wear ≥ 0.2 mm
- Nose wear ≥ 0.3 mm
- Notching at the depth of cutline ≥ 0.6 mm
- Surface roughness value ≥ 3.0 μm
- Excessive chipping or catastrophic failure

Figure 5.12 represents a comparison of surface roughness for dry and air-cooled turning for eight different cutting conditions. It can be noticed that the R_a values increases with cutting time irrespective of the cutting conditions. In general, the surface finish for air-cooled turning is found slightly better than dry turning. However, at a combination of high feed and low depth of cut (Fig. 5.12c,g), the surface finish for dry turning is found better compared to air-cooled turning.

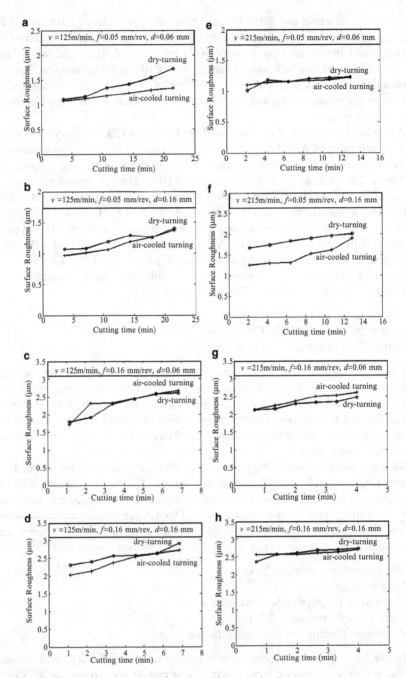

Fig. 5.12 Surface roughness variation with cutting time in turning of hardened steel with CBN tool (Sarma 2009)

Improvement in the surface finish for these conditions may be attributed to the change in cutting tool geometry. For the conditions pertaining to Fig. 5.12c, i.e., at low speed, high feed and low depth of cut, a built up edge (BUE) was formed near the chamfered edge of the insert during air-cooled turning. The BUE was not observed at high speed. As expected, the R_a value at lower feed rate (0.05 mm/rev) is found better than the R_a value at higher feed rate (0.16 mm/rev). Figure 5.12a, b, e, f corresponds to low feed rate with R_a values ranging from 1 to 2 μm, and Fig. 5.12c, d, g, h corresponds to the high feed rate with R_a values ranging from 1.7 to 2.8 μm. Generally, the surface finish in machining operations improves with increasing cutting speed. However, at conditions corresponding to Fig. 5.12f, the surface finish is found slightly higher compared to the condition at low speed (Fig. 5.12b). At this cutting condition (215 m/min, 0.05 mm/rev, 0.16 mm), nose wear occurred for both dry and air-cooled turning. The presence of nose wear deteriorates the surface finish during the turning operation.

Figure 5.13 represents a comparison of propagation of flank wear with cutting time for dry and air-cooled turning for eight different cutting conditions. In each condition, it has been observed that tool flank wear is higher during dry turning compared to air-cooled turning. This shows that air-cooling has a direct influence in reducing tool flank wear and thus increasing the tool life. The tool rejection criteria based on maximum flank wear was kept as 0.2 mm, but in actual turning, after the flank wear of about 0.07 mm, the surface roughness of the machined surface came out to be more than 3.0 μm and the tool was rejected based on maximum surface criterion. For the condition of high cutting speed (Fig. 5.13e–h), the tool flank wear is found greater compared to the low cutting speed (Fig. 5.13a–d). During high speed machining, temperature becomes high causing partial burning near the cutting edge and sometimes crater wear in the rake face. Burning of cutting edge accelerates the tool flank wear rate and deteriorates the surface finish. It has been observed that at high cutting speed, high feed, and high depth of cut (Fig. 5.13h), flank wear as well as crater wear was formed during dry turning but not in air-cooled turning. On the other hand, at high speed and low feed (Fig. 5.13e, f), flank wear as well as nose wear was seen during dry turning, but not in air-cooled turning. The experimental results show that air cooling may be used as a preventive measure to reduce the wear due to thermal effects.

Figures 5.14 and 5.15 represent a comparison of feed force (F_x) and cutting force (F_z) for dry and air-cooled turning for eight different cutting conditions. In general, the values of cutting force are greater than the feed force for the same cutting condition. In another observation, feed force and cutting force are found to be greater during air-cooled turning except for the conditions pertaining to Figs. 5.14e, f and 5.15e, f. The reason is that during air-cooled turning, air carries away the heat through convective heat transfer method and reduces the temperature. Therefore, the mechanical strength of the job material becomes high in comparison to dry turning due to which more feed/cutting force is required to do machining in air-cooled turning. However, for conditions of Figs. 5.14e, f and 5.15e, f, nose wear occurred along with flank wear during dry turning. It is generally believed that the forces are lower when machining at higher cutting speed. However, here, in some

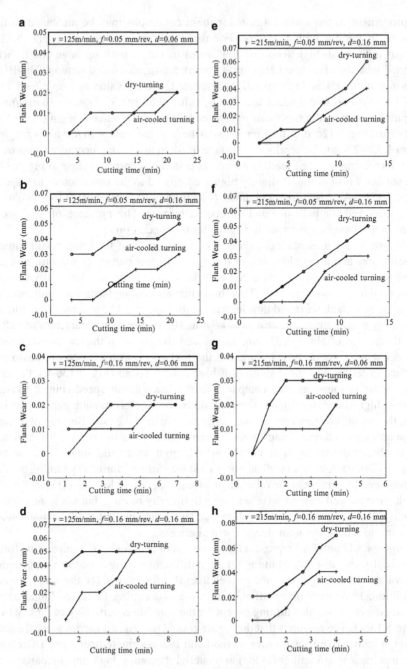

Fig. 5.13 Flank wear progression with cutting time in turning of hardened steel with CBN tool (Sarma 2009)

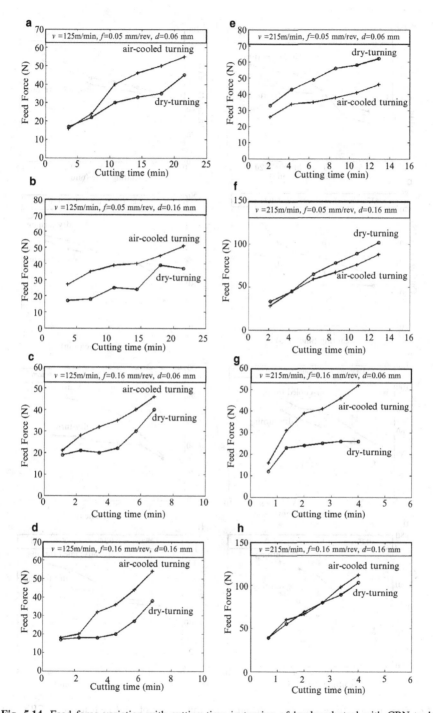

Fig. 5.14 Feed force variation with cutting time in turning of hardened steel with CBN tool (Sarma 2009)

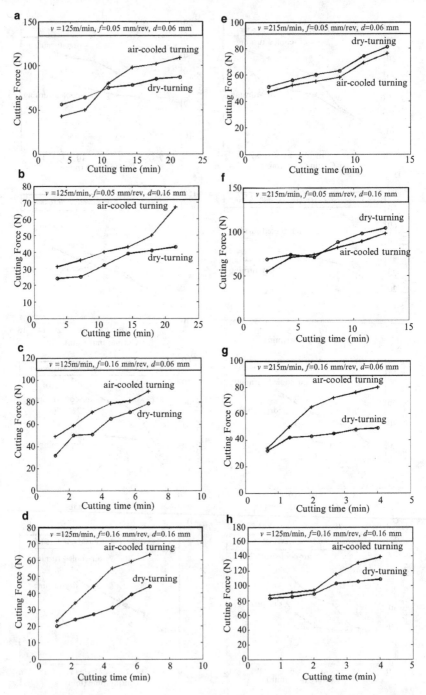

Fig. 5.15 Cutting force variation with cutting time in turning of hardened steel with CBN tool (Sarma 2009)

Fig. 5.16 Tool wear progression during (**a**) dry turning and (**b**) air-cooled turning of hardened steel with CBN tool (Sarma 2009)

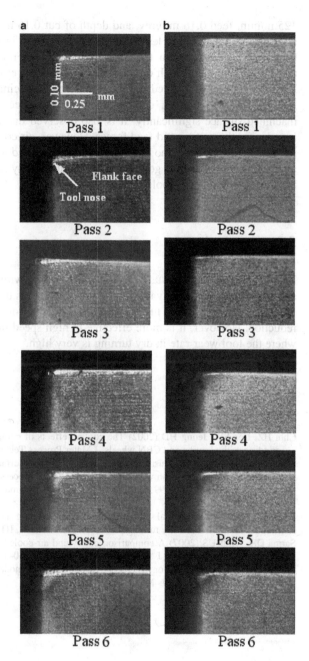

cases, the forces are found greater in high cutting speed. In these cases, besides flank wear, the nose wear or BUE was observed.

Figure 5.16 shows the images of progression of tool flank wear in dry and air-cooled turning of H13 steel with CBN tool for a condition of cutting speed

125 m/min, feed 0.16 mm/rev, and depth of cut 0.16 mm. Each image was taken after machining a cutting length of 100 mm. As seen in the images, the wear land keeps on increasing with time at almost constant rate. Comparing Fig. 5.16a and Fig. 5.16b, the growth of wear land is found lower during air-cooled turning. Thus, air cooling has played an important role in reducing the growth of wear land.

It is observed that air cooling does not influence surface roughness of the machined surface significantly. It is also observed that air cooling does not help in reducing the cutting and feed forces. However, air cooling significantly reduces the tool wear. Moreover, no built-up edge was observed with air cooling. Therefore, air-cooling seems to be a good environmental friendly and inexpensive option for hard turning with CBN tools.

5.5 Conclusion

This chapter presents a study of air-cooled turning with coated carbide, ceramic, and CBN tools. Readily available compressed air was used without any treatment. In all the cases, air-cooled turning provided better performance, particularly in wear reduction. However, it is more effective in high speed machining or hard turning, where the tool wear rate in dry turning is very high.

References

Choi HZ, Lee SW, Jeong HD (2002) The cooling effects of compressed cold air in cylindrical grinding with alumina and CBN wheels. J Mater Process Technol 127:155–158

Ghani AK, Choudhary IA, Husni (2002) Study of tool life, surface roughness and vibration in machining nodular cast iron with ceramic tool. J Mater Process Technol 127:17–22

Sadasivan TA, Sarathy D (1999) Cutting tools for productive machining, 1st edn. Widia (India) Limited, Bangalore

Sarma DK (2009) Experimental study, neural network modeling and optimization of environment-friendly air-cooled and dry turning processes. Ph.D. Thesis, IIT Guwahati

Sarma DK, Dixit US (2007) A comparison of dry and air-cooled turning of grey cast iron with mixed oxide ceramic tool. J Mater Process Technol 190:160–172

Smithy DW, Kapoor SG, DeVor RE (2000) A worn tool force model for three dimensional cutting operation. Int J Mach Tools Manuf 40:1929–1950

Chapter 6
Offline and Online Optimization
of Machining Processes

Abstract Optimization of machining processes enhances the machining performance, and it is possible to carry out efficient and economic machining operations without harming the environment. In this chapter, procedure for offline and online optimization is outlined. The methods to enhance the utilization of cutting tools in an environment-friendly manner are discussed.

Keywords Air-cooled turning • Cost of machining • Dry turning • Effective utilization of cutting tools • Finish machining • Flank wear • Multi-pass machining • Offline optimization • Online optimization • Optimization • Pareto-optimal • Weibull distribution

6.1 Introduction

Optimization of machining processes is of paramount importance. Properly optimized dry machining can give better performance than non-optimized machining with cutting fluid. Optimization can be carried out offline or online. In offline optimization, the machining parameters are chosen a priori based on the past experience. Online optimization is a posteriori optimization. Here, the machining parameters are changed based on the feedback of the process.

There are some differences in the perception of word "online." Some consider online optimization as an automatic and continuous adjustment of the cutting parameters without human intervention and process interruption for obtaining the best possible machining performance. In this chapter, the word "online optimization" has been used in a broader sense to mean the optimization by a human or machine operator in real-time (i.e., during actual production) by taking the shop floor feedback either through sensors or human sense organs. Thus, online optimization need not be automatic. It is different from offline optimization in the sense that offline optimization uses machining process models, like cutting force model, tool wear model, surface roughness model, etc. based on a priori knowledge

U.S. Dixit et al., *Environmentally Friendly Machining*, SpringerBriefs
in Applied Sciences and Technology, DOI 10.1007/978-1-4614-2308-9_6,
© Springer Science+Business Media, LLC 2012

gathered from offline experiments. This view point about offline optimization is also expressed by Park and Kim (1998).

The optimization of machining process is an old but active research area, with the pioneering work of Gilbert (1950). Since then thousands of papers have been published in this area. There are a number of review papers available on this topic (Aggarwal and Singh 2005; Mukherjee and Ray 2006; Ojha et al. 2009; Chandrasekaran et al. 2010; Dhavamani and Alwarsamy 2011). Almost all of the researchers have omitted the cost of cutting fluid in their optimization models. Most of the researchers have used a tool life equation considering the failure of tool due to wear. In actual practice, in finish turning a tool is discarded if it no longer provides the desired surface finish. However, the surface finish provided by the tool is not the function of tool condition alone. It is influenced by the cutting parameters (cutting speed, feed, and depth of cut) and cutting environment (type and method of applying cutting fluid), which need to be optimized for obtaining the best out of a cutting tool.

In this chapter, the procedure for offline and online optimization of machining process is outlined with emphasis toward cutting fluid and environmental concerns. A methodology for effective utilization of cutting tools is also discussed. The effective utilization of cutting tools has direct bearing on energy and environment.

6.2   Offline Optimization

The objectives of metal cutting problems are minimization of cost of machining, maximization of the production rate, and maximization of the profit rate. In general, all the objectives cannot be satisfied simultaneously. The optimization problem becomes a single-objective if one objective is chosen. If more than one objectives are chosen, it becomes a multiobjective problem. There are several ways of tackling multiobjective problems. The most popular and simple approach is to take a weighted average of various objective functions. Assignment of suitable weights corresponding to objective functions is a difficult task involving subjective assessment. Another way of solving a multiobjective problem is to obtain various Pareto-optimal solutions. Two solutions are called Pareto-optimal if no solution dominates the other in all aspects. In some aspects, one solution may be better and in some other aspects the other solution may be better. One can obtain a set of Pareto-optimal solutions, in which no solution can be said to be better than the other considering all objectives of the problem. Among this set, one solution can be chosen based on some higher level decision criterion.

In this section, the objective functions and constraints of the machining optimization are described considering the cost associated with cutting fluid. If the sufficient data about the machining behavior is available, the optimization problem can be solved to provide a proper solution. Considering the cost associated with the cutting fluid, in some cases, it may be possible to fulfill the desired objectives in dry machining or environmentally friendly cooling.

6.2.1 Objective Functions

In a multipass machining operation, the total production time per component, T_p, is expressed as

$$T_p = T_{tR} + \frac{t_c T_{tR}}{T_R} + T_{tF} + \frac{t_c T_{tF}}{T_F} + T_L + t_{ts}, \tag{6.1}$$

where T_{tR} is the total cutting time of rough machining, t_c the time required for changing a tool, T_R the tool life for rough machining, T_{tF} the total cutting time of finish machining, T_F the tool life for finish machining, T_L the loading and unloading time, and t_{ts} the total tool setting time. Total cutting time for rough machining is obtained as the summation of cutting times for m roughing passes. The total tool setting time for m roughing and one finishing pass is given by

$$t_{ts} = (m+1)t_s, \tag{6.2}$$

where t_s is the setting time for each pass.

A rearrangement of (6.1) provides

$$Tp = T_L + t_{ts} + T_{tR} + T_{tF} + t_c \left(\frac{T_{tR}}{T_R} + \frac{T_{tF}}{T_F} \right). \tag{6.3}$$

The first term in (6.3) is independent of the process parameters. The second term is dependent on the number of passes and should be considered in a multipass machining process. The objective of the machining process is to minimize T_p.

The cost of machining per piece is given as

$$F_c = C_o T_p + C_t \left(\frac{T_{tR}}{T_R} + \frac{T_{tF}}{T_F} \right) + C_F T_p, \tag{6.4}$$

where C_o is the operating cost per unit time that includes labor and overhead costs, C_t is the cost associated with tool change that includes the cost of one cutting edge of the tool and labor and overhead costs involved in changing the tool, and C_F is the per unit time cost associated with the cutting fluid that includes storage and disposal cost apart from the cost of cutting fluid itself. At this stage, it is assumed that C_o, C_t, and C_F are known. In Chap. 7, some detailed discussion regarding the calculation of these costs will be presented. In (6.4), cost associated with tool change is considered to be equal for roughing and finishing.

Substituting the value of T_p from (6.3) into (6.4),

$$F_c = (C_o + C_F) \left[T_L + t_{ts} + T_{tR} + T_{tF} + t_c^* \left(\frac{T_{tR}}{T_R} + \frac{T_{tF}}{T_F} \right) \right], \tag{6.5}$$

where the effective tool change time is given by

$$t_c^* = t_c + \frac{C_t}{(C_o + C_F)}. \tag{6.6}$$

As C_o and C_F are constant values independent of the process parameter, the objective function for the minimum production time per piece (6.3) can be converted into the objective function for minimum machining cost per piece by replacing t_c by t_c^*. The two objectives are close to each other if $C_t/(C_o+C_F)$ is small. One can obtain a number of solutions by solving the optimization problems of minimizing the production time by taking various values of t_c starting from the actual tool changing time to actual tool changing time plus $C_t/(C_o+C_F)$. All these solutions will be Pareto-optimal solutions. A higher level decision can be taken to choose the best solution among various Pareto-optimal solutions.

If the raw material cost of the workpiece is C_m and the price after machining is P_p, then the profit rate, P_R, is expressed as

$$P_R = \frac{P_p - F_c - C_m}{T_p}. \tag{6.7}$$

The solution for maximum profit rate minimizes the cost and production time objectives in a weighted average sense. Hence, it will be included in the Pareto-optimal set obtained by minimizing the objective functions of cost and production time.

It is clear that the objective functions can be evaluated only if tool life can be predicted. The tool life is dependent on cutting parameters and cutting environment i.e., the status of cutting fluid application. One of the most popular equations in turning process is Taylor's extended tool life equation given as

$$Tv^p f^q d^r = C, \tag{6.8}$$

where v is the cutting speed in m/min; T is the tool life in minutes; d is the depth of cut in mm; and p, q, r, and C are the constants dependent on a particular tool and work material combination as well as on the status of cutting fluid application. Determination of these constants is time consuming and expensive. Moreover, this equation is applicable over a narrow range of process parameters. Hence, it is important to have a reasonable guess for process parameters, so that the machining experiments can be concentrated around those process parameters.

Ojha and Dixit (2005) fitted a neural network model for obtaining the tool life as a function of cutting parameters (feed, speed, and depth of cut). Training a neural network requires a large number of data. To accelerate the time of training, the following procedure was adapted. The wear time curves for most cutting tools follow a pattern as shown in Fig. 6.1, having three distinct wear zones. These are initial wear zone, steady wear zone, and severe wear zone.

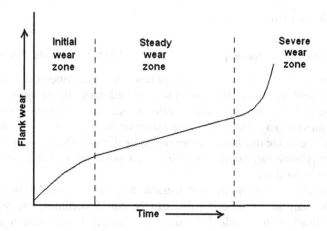

Fig. 6.1 Flank wear vs. cutting time for a typical tool

Tool life is estimated by fitting a best-fit line on the data falling in the steady wear zone. Thus,

$$w = a + bt, \tag{6.9}$$

where w is the flank wear, t is the time, and a and b are constants. The tool life T is given by

$$T = \frac{w_{max} - a}{b} \tag{6.10}$$

where w_{max} is the maximum flank wear. By conducting a few tests till tool failure, it was verified that wear-time curve was almost linear up to maximum flank wear of 0.8 mm, thus justifying the use of linear interpolation based on the best fit line.

In milling process also, empirical equations for tool life have been developed. For example, Tolouei-Rad and Bhidendi (1997) proposed an empirical relation to evaluate tool life T for milling process as

$$T = \frac{60}{Q} \left(\frac{C(0.2d/f)^g}{(df)^w v} \right)^{1/n}, \tag{6.11}$$

where T is the tool life in minutes; v is the cutting speed in m/min; C is a constant; f is the feed in mm/rev; d is the depth of cut in mm; and g, w, and n are exponents for different tool and work material combination. The parameter Q is contact proportion of machining time during which cutting edge is engaged with workpiece.

6.2.2   Constraints

The machining optimization problem is subjected to the following constraints:

Tool life constraint: Tool life should not be less than a prescribed value to avoid the frequent tool changes. The minimum desired tool life is generally based on subjective decision. As a thumb rule, the minimum desired tool life should be 20 times the machining time of one component. This is to ensure that at least 20 components are machined before replacing the tool. One can put a constraint on the maximum value of the tool life as well, but in most of the cases, it will be an inactive constraint.

Surface roughness constraint: A constraint that surface roughness should not be more than the prescribed value can be put. Sometimes, excessive better surface finish is not desired, because certain tribological and heat transfer characteristics are dependent on it. Therefore, the surface roughness value may be restricted to lie in a zone. For proper implementation of this constraint, a surface roughness prediction model, for example a neural network model, is needed.

Cutting force constraints: The components of cutting forces should be limited to avoid excessive job and tool deflection and breakage of the tool. The tool and job deflection and the stresses in the job can be found by physics based or soft computing based models.

Machine power constraint: The machine power can be calculated using the following formula:

$$\text{Machine power} = \frac{\text{main cutting force} \times \text{cutting speed}}{\text{efficiency of the machine}}. \qquad (6.12)$$

The machine power should be limited to avoid excessive overloading of the spindle motor. At the same time, if machine power is much lesser than the power of the spindle motor, the machine is underutilized.

Geometric constraint: There may be some restrictions based on the geometry of workpiece.

Temperature constraints: Temperature of the workpiece, machine tool, and cutting tool should not be excessive for dimensional accuracy as well as tool life. This constraint is especially important in dry machining.

Variable bounds: Cutting speed, feed, and depth of cut should lie within certain ranges. These ranges are dependent on the type of machine, type of tool, and type of material.

In addition to these constraints, there is a constraint that the number of passes m is an integer quantity.

6.3 Online Optimization

Nowadays, a number of tool–work combinations are used for turning in industries. A number of techniques have been proposed for optimizing the process parameters. However, almost all of them require the knowledge of tool life as a function of cutting parameters. It is not economical to conduct tool life tests for each combination. Sarma and Dixit (2009) proposed a heuristic based method for optimizing the finish turning process. The method does not require a priori information of tool life.

The objective function to be minimized in finish turning process is

$$\text{Minimize } F = \frac{1}{fv}\left(1 + \frac{t_c^*}{T_F}\right), \tag{6.13}$$

Assume that the upper bound of fv based on the constraints is x_1 and corresponding tool life is T_1. If any other fv is denoted by x_2 and the corresponding maximum possible tool life be denoted by T_2, then the following condition will yield a smaller value of the objective function (F) compared to the highest possible fv (Basak et al. 2007):

$$\frac{1}{x_2}\left(1 + \frac{t_c^*}{T_2}\right) < \frac{1}{x_1}\left(1 + \frac{t_c^*}{T_1}\right). \tag{6.14}$$

Thus, to economize the turning process fv can be reduced to x_2 by ensuring that

$$t_c^* > \frac{T_1 T_2 (x_1 - x_2)}{T_2 x_2 - T_1 x_1}, \tag{6.15}$$

Based on this the following heuristic algorithm is developed.

Step 1. Select the maximum possible feed based on the surface roughness consideration. Then select the maximum possible speed based on the maximum power available in machine tool and the bound on the cutting speed. Carry out the machining till the tool fails. Record the tool life T_1.

Step 2. Reduce the speed by 10% and carry out the machining till tool failure. Record the tool life T_2.

Step 3. For a small range, the tool life may be considered to vary in a linear way. Thus, consider that when fv is reduced by a factor of $(1 - c)$, the tool life T_c will be given by

$$T_c = (1 - 10c)T_1 + 10cT_2 \tag{6.16}$$

It is seen that for $c = 0, T_c = T_1$ and for $c = 0.1, T_c = T_2$. At optimum condition [using (6.15)],

$$t_c^* = \frac{cT_1T_c}{T_c(1-c) - T_1}.\tag{6.17}$$

The optimum value of c can be obtained by solving (6.17).

Step 4. Now the machining is carried out at a cutting speed of $(1 - c)$ times the maximum speed. The refined value of T_c can be obtained after the tool fails.

Step 5. If the actual T_c is same as that estimated by (6.16), the optimum is reached. If actual T_c is different than that obtained by (6.16), a new tool life relation is developed with obtained data and the procedure is repeated.

As an example, initially the dry turning was carried out at 300 m/min. The mean tool life T_1 corresponding to that speed is 5.63 min. When the cutting speed was reduced by 10%, the mean tool-life became 6.51 min. Assuming $t = t_c^*$ as 6 min, the value of c in this case is 0.12. Thus, the optimum speed is $(1-0.12) \times 300 = 264$ m/min. The machining may be carried out at this speed.

As more number of pieces is machined, information about tool life gets updated. With revised information, the optimization procedure may be revisited. Thus, in the presence of statistical variations, the optimization is a dynamic process initially. When sufficient number of data is available in the vicinity of expected optimum solution, the optimum solution may be frozen.

In this algorithm, only the speed is varied. One can easily develop an algorithm in which speed and feed can be varied simultaneously. A suitable optimization method such as simplex search can be used to optimize the process parameters.

6.4   Effective Utilization of Cutting Tools

Sarma and Dixit (2009) proposed methods for effective utilization of tools in dry and air-cooled turning. The methods were tested by conducting a number of experiments on finish turning of steel with coated carbide tools. A cutting tool was considered failed if it no longer provided a surface roughness value less than 2.5 μm. The methods are discussed in the following subsections.

6.4.1   Enhancing the Tool Life with Air-Cooling

It was discussed in Chap. 5 that obtained surface finish is slightly better in dry turning compared to air-cooled turning, but the flank wear is significantly reduced in air-cooled turning. Thus, the tool life based on surface roughness is lower for air-cooled turning, although the flank wear is reduced. After getting failed in air-cooled turning, the tools can be reused in dry turning till they fail in dry turning too. To assess this strategy, seven experiments at each of the cutting speed of 270 m/min and 300 m/min were carried out. Each tool was first used in air-cooled turning. When surface roughness of machined components exceeded the prescribed limit of 2.5 μm, air flow was stopped and dry turning was carried out. The dry turning continued till the surface roughness exceeded 2.5 μm.

The probability density function (by fitting Weibull distribution) of the tool life is plotted in Fig. 6.2, Fig. 6.2a corresponding to cutting speed of 300 m/min, and Fig. 6.2b corresponding to tool life of 270 m/min. Table 6.1 shows the mean tool life and corresponding maximum flank wear. For comparison, Table 6.2 is also reproduced from Sarma and Dixit (2009) that shows the tool lives in dry and air-cooled turning. A comparison of two tables reveals that the turning carried out with this strategy enhances tool life. In the present case, tool life got increased by 42% at cutting speed of 300 m/min and 38% at cutting speed of 270 m/min.

An explanation for the success of the present strategy is as follows. Initially, the rate of wear is high and air cooling is quite effective in reducing the flank wear. Till some time, cutting edge remains sharp and provides a good surface finish. After some time, the change in cutting edge and tool geometry due to wear deteriorates the surface roughness. At this stage, dry turning is carried out, in which the material softening due to temperature compensates for the deterioration of surface roughness due to tool wear.

Although the strategy is quite effective, it introduces the additional cost of compressed air. One can carry out the cost-benefit analysis for a specific situation. This is discussed in the next chapter devoted exclusively for economic aspects of machining.

6.4.2   Enhancing the Tool Life by Changing Cutting Conditions

A cutting tool should be used at the optimum cutting conditions till failure. If the tool failure is based on the maximum flank wear, the tool cannot be reused. However, if it is based on the maximum surface roughness, there is a possibility to reuse it by changing the cutting conditions. In fact, some numerical control machining strategies propose online tuning of cutting conditions to maximize material removal rate without violating the machining constraints (Merdol and Altintas 2008). In finish turning, if the feed is reduced the surface roughness reduces. Of course, too much reduction in the feed will be counter productive, as it reduces the production rate.

A simple way to assess whether reusing the tool at the reduced feed is economical is as follows. The cost of turning a workpiece of diameter D and length L is given by

$$C = C_0 \frac{\pi L D}{f v} \left[1 + \frac{t_c + C_t/C_0}{T} \right], \tag{6.18}$$

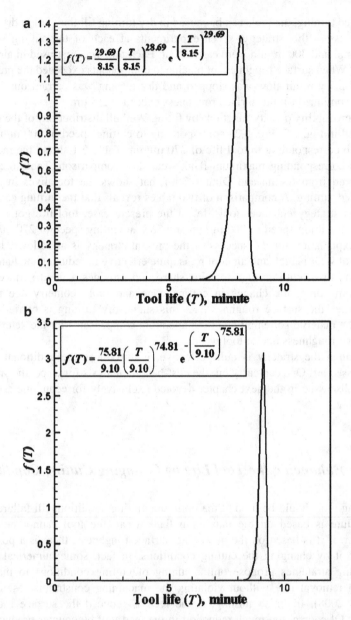

Fig. 6.2 Probability distributions of tool life when dry turning followed air-cooled turning (**a**) Cutting speed = 300 m/min (**b**) Cutting speed = 270 m/min. With permission from Sarma and Dixit (2009). Copyright (2009) ASME

Table 6.1 Mean tool lives and maximum flank wears when dry turning followed the air-cooled turning

Cutting speed (m/min)	Mean tool life (min)	Maximum flank wear (mm)	
		Average value	Standard deviation
300	8.00	0.13	0.0078
270	9.03	0.11	0.0115

With permission from Sarma and Dixit (2009). Copyright (2009) ASME

Table 6.2 Mean tool lives and maximum flank wears at different cutting conditions

Type of turning	Cutting speed (m/min)	Mean tool life (min)	Maximum flank wear (mm)	
			Average value	Standard deviation
Dry	300	5.63	0.15	0.0149
Dry	270	6.51	0.12	0.0095
Air-cooled	300	4.90	0.08	0.0078
Air-cooled	270	6.00	0.07	0.0069

With permission from Sarma and Dixit (2009). Copyright (2009) ASME

where C_0 is the operating cost per minute, C_t the tool cost per edge, and t_c is the tool change time. Assume that f_1 is the optimum feed and f_2 is the reduced feed after the tool failed based on the maximum surface roughness at feed f_1. Thus, the cost of turning the workpiece at feed f_1 is

$$C_1 = C_0 \frac{\pi L D}{f_1 v} \left[1 + \frac{t_c + C_t/C_0}{T_1} \right] \tag{6.19}$$

and the cost of turning the workpiece at feed f_2 is

$$C_2 = C_0 \frac{\pi L D}{f_2 v} \left[1 + \frac{t_c}{T_2} \right]. \tag{6.20}$$

Note that in the expression for C_2, the tool cost is taken as zero. Reusing the tool at the reduced feed will be economical as long as C_2 is less than C_1. Thus, the required condition for economical machining is

$$C_0 \frac{\pi L D}{f_2 v} \left[1 + \frac{t_c}{T_2} \right] < C_0 \frac{\pi L D}{f_1 v} \left[1 + \frac{t_c + C_t/C_0}{T_1} \right] \tag{6.21}$$

or

$$f_2 > \frac{f_1 \left[1 + \frac{t_c}{T_2} \right]}{\left[1 + \frac{t_c + C_t/C_0}{T_1} \right]}. \tag{6.22}$$

Table 6.3 Total tool lives and maximum flank wears when tool was first operated at 0.1 mm/rev and then at 0.07 mm/rev in dry turning

Cutting speed (m/min)	Mean tool life (min)	Maximum flank wear (mm)	
		Average value	Standard deviation
300	10.82	0.17	0.023
270	10.97	0.14	0.012

With permission from Sarma and Dixit (2009). Copyright (2009) ASME

If the tool change time is a very small fraction of tool life, then a simplified relation is

$$f_2 > \frac{f_1}{\left[1 + \frac{C_t/C_0}{T_1}\right]}. \tag{6.23}$$

Taking $C_0 = \$0.06/\text{min}$, $C_t = \$2/\text{min}$, $T_1 = 6$ min, and $f_1 = 0.1$ mm/rev, the following condition is obtained: $f_2 > 0.015$ mm/rev.

For experimental verification of this strategy, after the tool failed at a feed of 0.1 mm/rev in dry turning, it was operated at the feed of 0.07 mm/rev. Table 6.3 (when compared with Table 6.1) shows that in this manner the tool life got enhanced by 68% at 270 m/min and by 92% at 300 m/min. Table 6.3 also shows that although the flank wear keeps increasing, it was still less than 0.6 mm, when the tool failed based on the surface roughness at a feed of 0.07 mm/rev. Figure 6.3 shows the probability distribution of the total tool life at dry turning conditions.

6.5   Conclusion

Optimization of the machining processes is required for saving the energy and environment. Although a lot of work has been done in the area of optimization of machining processes, consideration of costs associated with the cutting fluid is lacking in most of the work. It is possible to use the power of process optimization to eliminate harmful cutting fluids. For example, based on a lot of experiments on finish turning of steel using coated carbide inserts, Diniz and Micaroni (2002) observed that by increasing feed and tool nose radius and decreasing speed, dry turning can provide almost the same performance as the wet turning. Further, dry cutting requires less power and produces a smoother surface than wet cutting. As discussed in this chapter, Sarma and Dixit (2009) have suggested ways of enhancing the tool utilization using air cooling. In the next chapter, economics of environmentally friendly machining will be discussed.

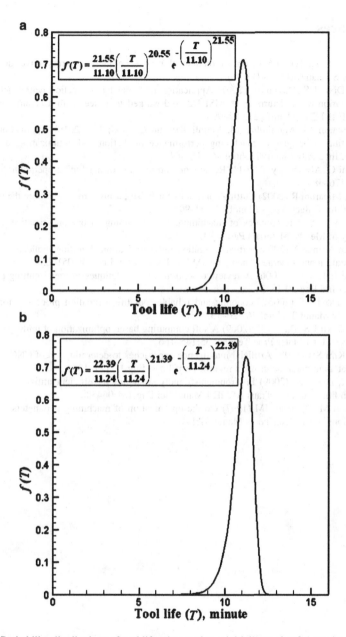

Fig. 6.3 Probability distributions of tool life when tool was initially used at 0.1 mm/rev feed and after failure at this feed, the feed was reduced to 0.07 mm/rev (**a**) Cutting speed = 300 m/min (**b**) Cutting speed = 270 m/min. With permission from Sarma and Dixit (2009). Copyright (2009) ASME

References

Aggarwal A, Singh H (2005) Optimization of machining techniques—a retrospective and literature review. Sadhana 30:699–711

Basak S, Dixit US, Davim JP (2007) Application of radial basis function neural networks in optimization of hard turning of AISI D2 cold-worked tool steel with a ceramic tool. Proc IMechE B J Eng Manuf 221:987–998

Chandrasekaran M, Muralidhar M, Murali Krishna C, Dixit US (2010) Application of soft computing techniques in machining performance prediction and optimization: a literature review. Int J Adv Manuf Technol 46:445–464

Dhavamani C, Alwarsamy T (2011) Review on optimization of machining operation. Int J Acad Res 3:476–485

Diniz AE, Micaroni R (2002) Cutting conditions for finish turning process aiming: the use of dry cutting. Int J Mach Tools Manuf 42:899–904

Gilbert WW (1950) Economics of machining. In: Machining theory and practice. American Society of Metals, Materials Park, OH, pp. 465–485

Merdol SD, Altintas Y (2008) Virtual simulation and optimization of milling applications—part II: optimization and federate scheduling. ASME J Manuf Sci Eng 130:051005

Mukherjee I, Ray PK (2006) A review of optimization techniques in metal cutting processes. Comput Ind Eng 50:15–34

Ojha DK, Dixit US (2005) Economic and reliable tool life estimation procedure for turning. Int J Adv Manuf Technol 26:726–732

Ojha DK, Dixit US, Davim JP (2009) A soft computing based optimization of multi-pass turning processes. Int J Mater Prod Technol 35:145–166

Park KS, Kim SH (1998) Artificial intelligence approaches to determination of CNC machining parameters in manufacturing: a review. Artif Intell Eng 12:127–134

Sarma DK, Dixit US (2009) Environment-friendly strategies for efficient utilization of cutting tools in finish turning. Trans ASME J Manuf Sci Eng 131:064506

Tolouei-Rad M, Bhidendi IM (1997) On the optimization of machining parameters for milling operations. Int J Mach Tools Manuf 37:1–16

Chapter 7
Economics of Environmentally Friendly Machining

Abstract In this chapter, economics of environmentally friendly machining is discussed. It is emphasized that in the machining cost calculation, the costs associated with cutting fluid storage and disposal should be considered. At the same time, the impact on environment can also be converted in a cost form. A method to assess the cost of air-cooled machining is presented, and it is shown that typically air-cooled machining is a viable option both economically and environmentally.

Keywords Air-cooled turning • Cost calculation • Cost of machining • Depreciation cost • Direct cost • Dry turning • Economics • Environmentally friendly machining • Indirect cost

7.1 Introduction

Environmental concerns are very important and must be followed religiously. However, a manufacturing industry is driven by profit motive and a technological solution is difficult to implement unless it is profitable in monetary form. Fortunately, in many situations, environmentally friendly machining is economical also, provided proper accounting of costs involved is carried out. In this chapter, cost calculation methodology is described considering the cost associated with cutting fluid. It is also shown that simple air-cooled machining is economical compared to dry machining in many situations.

U.S. Dixit et al., *Environmentally Friendly Machining*, SpringerBriefs
in Applied Sciences and Technology, DOI 10.1007/978-1-4614-2308-9_7,
© Springer Science+Business Media, LLC 2012

7.2 Determination of Cost of Machining

In Chap. 6, the expression for cost of machining per piece was provided as

$$F_c = C_o T_p + C_t \left(\frac{T_{tR}}{T_R} + \frac{T_{tF}}{T_F} \right) + C_F T_p. \tag{7.1}$$

The precise determination of C_o, C_t, and C_F is not an easy task and requires some skill. Some elaboration on the determination of these cost components is provided. The operating cost C_o can be expressed as

$$C_o = C_d + C_i, \tag{7.2}$$

where C_d is the direct cost and C_i is the indirect cost. The major component of direct cost are the direct labor cost and electricity. The cost of electricity is usually small and is neglected. Indirect cost will include managerial overhead, indirect labor, rent for the space occupied by machine tool, depreciation, and maintenance. There are various ways to calculate depreciation cost of the machine tool. The simplest is the straight line depreciation method. Here, the purchase price of the machine tool is divided over its useful life. Thus, if a machine tool has been purchased at $100,000 and its useful life is 5 years, then depreciation cost per annum will be $100,000/ 5 = $20,000. This is based on the assumption that salvage value of the machine tool after 5 years is zero. If the salvage value of the product is not zero after its use, then for calculating annual depreciation cost, the purchase price is subtracted by salvage value. Thus, if after 5 years, machine tool or its parts can be disposed of at $20,000, then annual depreciation cost will be ($100,000-$20,000)/5 = $16,000. There are many other, more popular, and precise methods of calculating depreciation like reducing balance method, in which annual depreciation cost is not constant throughout its useful life but keeps decreasing with the passage of time.

The maintenance cost can be fixed at 10% the price of machine tool. The machining time includes the time when no cutting is taking place such as during over travel or moving of the tool from one location to other location. Inspection cost can also be included. Bank interest on the investment can also be included as a cost.

The cost associated with the cutting tool includes the cost incurred in changing the tool and the tool cost itself. Tool cost depends on the tool being a disposal insert or a regrindable tool. In the disposal insert, the cost can be divided between cutting edges. The tool holder cost can be added in the cost of tool. Usually, the tool holder life is 500 times the tool life. Thus, the cost of tool holder can be distributed on 500 cutting edges.

Cost of cutting fluid including its disposal and storage cost are denoted by C_F. Considering that the flow rate of coolant/cutting fluid is constant during machining, the cost can be expressed as linearly proportional to production time. This is a more conservative approach than taking the cutting fluid cost as proportional to only machining time. During the idle period of the tool, usually there is no flow of the coolant, but the storage cost is operative. Thus, attempt should be to minimize the production time for reducing the cutting fluid related cost. The best thing is to eliminate the cutting fluid without affecting the machining performance.

Cauchick-Miguel and Coppini (1996) have suggested a simplified alternative method of machining cost determination. Recently, Branker et al. (2011) have proposed a microeconomic model that can optimize machining parameters and can include all energy and environment costs. In this model, cost associated with CO_2 production has been considered. Authors show that when environmental or CO_2 costs are added, an apparently cheaper production facility may come out to be expensive.

7.3 Economics of Air-Cooled Turning

In Chap. 6, a strategy for efficient utilization of the tool in finish turning was proposed. According to it, first the air-cooled machining is carried out. Once, the tool is unable to generate required surface finish, the dry turning is employed. Dry turning improves the surface finish and it is carried out till the cutting is unable to generate desired surface finish even in dry turning. Although the strategy is quite effective, it introduces the additional cost of compressed air. If the increase of tool life due to air cooling is significant, then the use of air would come out to be economical. A simplified mathematical analysis to assess the economic benefit is as follows (Sarma 2009).

Let the tool edge cost be C_t and total tool life in dry turning be denoted as T_1. Then, the cost of tool edge per unit time in dry turning is equal to C_t/T_1. Assume the total tool life for both air-cooled and dry turning is T_2. If C_a is the cost of compressed air per minute, the total cost of compressed air is C_aT_3, where T_3 is the tool life in air-cooled turning. The total cost of the tool and air is $(C_t + C_aT_3)$. Therefore,

$$\text{Cost of tool edge and air per unit time} = \frac{C_t + C_aT_3}{T_2}. \tag{7.3}$$

The machining will be economical as long as the cost of tool edge and air per unit time is less than the tool edge cost per unit time in dry turning. Thus, the required condition for economical machining is

$$\frac{C_t + C_aT_3}{T_2} < \frac{C_t}{T_1} \tag{7.4}$$

or

$$\frac{C_aT_3}{T_2} < \left(\frac{C_t}{T_1} - \frac{C_t}{T_2}\right). \tag{7.5}$$

From the above equation, the critical value of cost of compressed air per minute is obtained as

$$(C_a)_{crit} = \frac{C_t}{T_1} \frac{(T_2 - T_1)}{T_3}. \tag{7.6}$$

If the actual cost of compressed air per minute is less than the critical value calculated from (7.6), then the proposed strategy of air-cooled turning followed by dry turning will be economical.

In a typical workshop at India, the compressed air cost was about $0.01/min. One edge of the cutting tool priced $2. The critical cost $(C_a)_{crit}$ of air per minute can be determined from (7.6) and using the mean tool life data. For cutting speed of 300 m/min, the critical cost of air is about $0.17/min (and for cutting speed of 270 m/min, it is about $0.13/min. The actual compressed air cost is $0.01 per minute. Hence, employing the air results in saving of the tool cost in addition to reducing the problem of worn tool disposal.

7.4   Conclusion

In this chapter, economics of environmentally friendly machining is discussed. It is emphasized that cost associated with the storage and disposal of cutting fluid should be considered in the calculation of cost of machining. Further, the impact on the environment and health of operator should also be considered and should be quantified in the form of cost. In many situations, employing the harmless coolant like air comes out to be economical apart from being an environmentally friendly alternative.

References

Branker K, Jeswiet J, Kim IY (2011) Greenhouse gases emitted in manufacturing a product—a new economic model. CIRP Ann Manuf Technol 60:53–56

Cauchick-Miguel PA, Coppini NL (1996) Cost per piece determination in machining process: an alternative approach. Int J Mach Tools Manuf 36:939–946

Sarma DK (2009) Experimental study, neural network modeling and optimization of environment-friendly air-cooled and dry turning processes. Ph.D. Thesis, IIT Guwahati

Chapter 8
Epilogue: Looking at the Future

Abstract This chapter concludes the monograph on environmentally friendly machining. A brief summary of topics covered has been presented in the Introduction section of the chapter. Five hot research areas are identified for contributing to environmentally friendly machining. They are development of cutting fluids, development of cutting tools, development of machine tools, optimization, and enhancing the machinability of the work material.

Keywords Biodegradability • Environmentally friendly machining • Minimal quantity lubrication • Oxidation stability • Self-lubricating tools • Solid lubricant • Storage stability • Wiper insert

8.1 Introduction

The main objective of this monograph is to create awareness about environmentally friendly machining. The traditional machining is a widely used manufacturing method and will continue to be a major manufacturing activity in decades to come. Although a number of nontraditional advanced manufacturing technologies have been developed, they are not competitive enough to replace traditional machining processes in many situations. Moreover, other advanced manufacturing technologies have their own environmental burdens and research has to be carried out to make them green. This monograph, however, concentrates on issues that make traditional machining a green process.

Two major factors affecting the environmental aspects of green machining are the cutting fluid and cutting tool. Cutting fluid can be in the form of liquid or gas and its purpose is to lubricate the tool–work interface and to cool machining zone. Additionally, cutting fluid helps to wash away the chips. Nowadays, some solid lubricants are being used in machining. Gopal and Rao (2004) used graphite as a solid lubricant in the grinding of SiC. The tangential force components and specific energy were reduced using graphite as a solid lubricant. Reddy and Rao (2006)

employed solid lubricants, graphite, and molybdenum disulfide in milling of AISI
1045 steel by solid coated carbide end mill cutters. Molybdenum disulfide provided
better machining performance. In general, solid lubricants are poor in cooling and
chip disposal. Moreover, one has to see if they are environmentally friendly.

The major focus of this monograph has been on cutting fluids in the form of
liquid and gas. A brief description of self-lubricating cutting tools is provided, but
these tools have yet to get popularity in shop floor. For this to happen, they have to
be manufactured at low cost in order to compete with existing tools. At the same
time, their ingredients should not be harmful to environment.

Although the work of a number of researchers has been referred in this monograph,
more details have been provided regarding the air-cooled machining carried out by
Sarma (2009) and few others. This method of machining is pollution free, economical, and technologically simple. It can be easily implemented by large and small
machine shops. A detailed comparison of air-cooled turning with dry turning has been
presented. The practicing engineers and researchers can easily verify these results and
optimize the process further by changing the air flow rate and angle of the nozzle.

For sustainable manufacturing, efficient utilization of cutting tool is also important
as the cutting tool manufacturing process put a burden on energy and environment.
The disposal of worn out tool is also of concern. It has been observed that in most
machine shops, cutting tools are underutilized. Efficient utilization of cutting tool not
only helps in reducing the cost of machining, but it also reduces the overall energy
consumption, thus helping the environment. In this monograph, some strategies for
the efficient utilization of cutting tool and machining process optimization have been
discussed.

For environmentally friendly machining, the research focus has to be in
developing cutting fluids, developing cutting tools, developing machine tools,
optimization, and enhancing the machinability of the materials. These are discussed
in the following sections.

8.2   Development of Environmentally Friendly Cutting Fluids

There have been a number of studies considering dry machining, minimal quantity
lubrication (MQL) machining, and flood coolant machining. There has been lesser
number of studies that compare various types of cutting fluids, from the point of
view of machining performance and environmental concerns. There is a need to
compare the available cutting fluids and at the same time to develop new cutting
fluids. Developing new cutting fluids requires the knowledge of metal cutting,
tribology, and chemistry.

Besides providing good machining performance, an environmentally friendly
cutting fluid should have the following properties:

1. Biodegradability: The cutting fluid should be biodegradable. The biodegradability
   can be measured from the loss of dissolved organic carbon in the cutting fluid
   in the presence of microorganism over a period of time.

2. Oxidation stability: MQL lubricants should be stable against thin film oxidation.
3. Storage stability: An MQL system consumes very little lubricant. Hence, the lubricant must remain stable for a long time. The storage ability can be measured by observing the changes in viscosity and total acid number.

Some recent developments in this area are summarized. As a result of rigorous experimental study, Suda et al. (2002) observed that polyol esters provide a very good overall performance in tapping operations. They synthesized polyol esters from a specific polyhydric alcohol rather than glycerin. Wakabayashi et al. (2003) observed that polyol esters provide a very good performance in turning process. The cutting performance of MQL machining with biodegradable esters was equivalent to or better than that of conventional machining with flood cutting fluid supply. Suda et al. (2004) proposed that a fully synthetic biodegradable polyol ester of viscosity grade 32 can be a candidate of base stock of various lubricants used in machine tool. This multifunctional fluid is expected to become a successful lubricant applicable to machining zone as well as other moving parts of the machine tool. Sharif et al. (2009) studied the possibility of using palm oil. Palm trees are in abundance in Malaysia, the home country of the authors. The authors observed that MQL system with palm oil provides a very good tool life and a satisfactory surface finish.

8.3 Development of Environmentally Friendly Cutting Tools

Cutting fluids are used to lubricate the tool–workpiece and remove the heat generated in machining zone. The burden on cutting fluids reduces if the cutting tools can be improved. One way is to develop self-lubricating tools. A self-lubricating tool reduces the friction at the tool work interface. Consequently, the cutting force and heat generation in the machining zone gets reduced. Other way is to develop cutting tools having high hot hardness. It will be possible to carry out dry machining with these tools.

As discussed in this monograph, it is possible to use tools with arrangement for internal cooling. In particular, the use of heat pipe in a cutting tool seems very interesting. Attempt should be made to make them cost-effective.

There have been some attempts to develop more effective cutting tools. Recently, the use of wiper inserts has shown much improved performance in turning. A typical wiper insert for turning is shown in Fig. 8.1. Instead of a single nose radius, this tool has a multiradii profile. The essence in hard turning with multiradii wiper tools is to give insert a more flat surface on the secondary cutting edge and wipe the scallops that are typically formed in inserts with a simple nose radius (Özel et al. 2007). This provides better surface finish even at high feed rates and dry machining conditions. El-Hossain (2010) has developed a two-point cutting tool for turning that can carry out roughing and finishing in the same pass. The author claims that his design of cutting tool provides a better chance for cutting fluid to be more effective in lubricating and cooling the primary cutting edge.

Fig. 8.1 A typical wiper
insert. From Özel et al.
(2007). Copyright (2007)
Elsevier

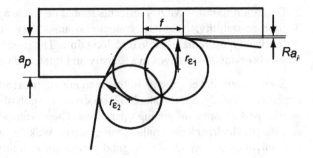

Fig. 8.1 A typical wiper insert. From Özel et al. (2007). Copyright (2007) Elsevier

8.4 Development of Environmentally Friendly Machine Tools

It has been emphasized that proper method of applying the coolant/cutting fluid jet is crucial to machining performance. Environmentally friendly machine tools should have proper arrangement for cutting fluid application. At present, most of the time, accessory for cutting fluid application, like MQL system, is retrofitted to machine tool, which does not provide optimum machining performance.

It is also possible to develop machine tools having small in-built compressor and a vortex tube for cooling the air. Further, CNC machine tools may be developed to have provision for appropriate thermal compensation. They may be provided with efficient chip disposal system. This will enable to carry out dry machining on the machine tools in an effective manner.

8.5 Optimization of Machining Processes Considering Environmental Impact

There has been a lot of research in the optimization of machining processes. Many a times, machining optimization problem has been used as a bench mark for comparing various optimization algorithms. It is rare to see the shop floor application of developed optimization strategies. This is mainly due to lack of reliable tool life and surface roughness models. Moreover, machining performance is also dependent on the condition of machine tool, making offline optimization solution far away from the actual optimal solution. There is a need to develop simple and effective optimization solutions that can be outlined by process planners and fine-tuned by machine operators.

It is essential to consider environmental aspects in optimization. At present, very little attention has been paid in this direction. It is often said that optimization is a philosophy rather than just a mathematical technique. Optimization considering environmental aspects will help to create awareness of the need to reduce environmental pollution.

8.6 Development of Environmentally-Friendly Workpiece Materials

One way of reducing or eliminating use of cutting fluid and enhancing the tool life is to increase the machinability of workpiece material. This has been attempted since long. For example, addition of a small amount of sulfur enhances the machinability of steel. Sulfur combines with steel to produce manganese sulfide (MnS) inclusion that acts as stress raiser and help in easy shearing during machining. Like this several metallurgical treatments can be employed to enhance the machinability of materials. Oxide treatment of free machining steels offers the possibility of forming selective transfer of built-up layers allowing better tool life at higher cutting speeds than their standard free machining counterparts (Hamann et al. 1996).

The area of developing materials with enhanced machinability will be hot in near future. However, it is a challenging task to develop the materials with enhanced machinability without sacrificing their other properties.

8.7 Conclusion

This monograph has covered some important issues of environmentally friendly machining. It is clear that a lot of research has been carried out in this field since the beginning of this century. Due to increasing world population and life style changes, the environmental pollution is steadily increasing. This is of major concern to every human being living on earth and everyone should try to cause minimum damage to environment in any activity. Machining is no exception. Its impact on environment is significant in developed and developing countries. There is a need to implement the developed green technologies in machine shops. At the same time, the research efforts should continue in the subareas discussed in this chapter.

References

El-Hossain EM (2010) Enhancement of surface quality using a newly developed technique in turning operations. Proc IMechE B J Eng Manuf 224:1389–1397

Gopal AV, Rao PV (2004) Performance improvement of grinding of SiC using graphite as a solid lubricant. Mater Manuf Process 19:177–186

Hamann JC, Grolleau V, Maitre FL (1996) Machinability improvement of steels at high cutting speeds—study of tool/work material interaction. CIRP Ann Manuf Technol 45:87–92

Özel T, Karpat Y, Figueira L, Davim JP (2007) Modelling of surface finish and tool flank wear in turning of AISI D2 steel with ceramic wiper inserts. J Mater Process Technol 189:192–198

Reddy NSK, Rao PV (2006) Experimental investigation to study the effect of solid lubricants on cutting forces and surface quality in end milling. Int J Mach Tools Manuf 46:189–198

Sarma DK (2009) Experimental study, Neural network modeling and optimization of environ-
    ment-friendly air-cooled and dry turning processes. Ph.D. Thesis, IIT Guwahati
Sharif S, Yusof NM, Idris MH, Ahmad ZA, Sudin I, Ripin A, Zin MAHM (2009) Feasibility study
    of using vegetable oil as a cutting lubricant through the use of minimum quantity lubrication
    during machining, Research Vot Number: 78055, Department of Manufacturing and Industrial
    Engineering, Faculty of Mechanical Engineering, Universiti Teknologi Malaysia
Suda S, Yokota H, Inasaki I, Wakabayashi T (2002) A synthetic ester as an optimal cutting fluid
    for minimal quantity lubrication machining. CIRP Ann Manuf Technol 51:95–98
Suda S, Wakabayashi T, Inasaki I, Yokota H (2004) Multifunctional application of a synthetic
    ester to machine tool lubrication based on MQL machining lubricants. CIRP Ann Manuf
    Technol 53:61–64
Wakabayashi T, Inasaki I, Suda S, Yokota H (2003) Tribological characteristics and cutting
    performance of lubricant esters for semi-dry machining. CIRP Ann Manuf Technol 52:61–64